Integrated Algebra Made Easy Handbook

2nd Edition

By:
Mary Ann Casey

B. S. Mathematics, M. S. Education

Acknowledgments

Thank you to my colleague and friend, Kimberly Knisell, who helped me get organized. Her proofreading and suggestions were invaluable as well. Also special thanks go to my publisher and editor, Keith Williams, and his assistant, Julieen Kane. Working with math symbols and diagrams is difficult even for a math teacher and these two people put it all together in an attractive and legible format for publication.

Dedication

To my former and future students — you make teaching a joy!

About The Author

The author has a B.S. in Mathematics and a M.S. in Education from the state University of New York at New Paltz. She has been teaching high school math for 19 years and has taught several courses at Ulster County Community College. Mrs. Casey is currently the lead math teacher in her high school, has served on committees at the NY State Education Dept., has been named in Who's Who Among American Teachers for the past 5 years, and maintains professional memberships in the National Council of Teachers of Mathematics and in the Association of Mathematics Teachers in New York State.

Introduction

Integrated Algebra Made Easy is a quick reference guide to the first year of high school algebra integrated with other topics which include statistics, probability, and the use of algebra in geometry. It is specifically coordinated with the New York State Curriculum for Integrated Algebra. In the interest of "good teaching of mathematics," some additional material that broadens the scope and understanding of the actual curriculum content is included in this handbook.

My approach to teaching is to make algebra "student friendly". Integrated Algebra Made Easy was developed from my own classroom teaching notes and my review materials for New York State Regents examinations which I aligned to the New York State curriculum. Examples of each procedure are provided but I have not included sets of practice problems. Practice problems are easily available in textbooks and by using Practice Tests For Integrated Algebra by Topical Review Book Company. This book provides a place to find out "How to do it!"

National and state standards require that the study of algebra today include extensive problem solving skills that are applied to real world problems. While many real world applications can be solved using other techniques, the tried and true algebraic method maintains its number one status as an efficient and effective tool for problem solving. Since so many kinds of real world problems exist, it isn't possible to address them all in a short reference guide. Each teacher, and each student, is encouraged to apply the algebraic method to as many different types of problems as they can develop.

A strong algebra foundation is required for students to continue successfully in high school and college academic math courses. It is also very helpful in dealing with everyday math situations. It is my hope that this review book will help all of our students establish a good rapport with Integrated Algebra – confidence, good skills, success, and an enjoyment of using algebra!

Sincerely,

MaryAnn Casey
Math Teacher
Saugerties High School

Table of Contents

CHAPTER 1: SETS AND SET NOTATION 1

CHAPTER 2: NUMBERS AND ALGEBRA 3

CHAPTER 3: PROPERTIES AND LAWS 7

CHAPTER 4: MATHEMATICAL SYSTEMS 10

CHAPTER 5: SIGNED NUMBERS 13

CHAPTER 6: MONOMIALS AND POLYNOMIALS 17

CHAPTER 7: EXPONENTS 22

CHAPTER 8: SCIENTIFIC NOTATION 24

CHAPTER 9: EVALUATING ALGEBRAIC EXPRESSIONS
AND FORMULAS 26

CHAPTER 10: FACTORING 28

CHAPTER 11: ALGEBRAIC FRACTIONS 34

CHAPTER 12: SIMPLIFYING RADICALS 41

CHAPTER 13: RATIO AND PROPORTION 44

CHAPTER 14: DIMENSIONAL ANALYSIS OR
CONVERSION FACTORS 50

CHAPTER 15: SIMPLE EQUATIONS 53

CHAPTER 16: SYSTEMS OF EQUATIONS (Simultaneous Equations) .. 58

CHAPTER 17: SOLVING SIMPLE INEQUALITIES 61

CHAPTER 18: COORDINATE GRAPHING 63

CHAPTER 19: SOLVING QUADRATIC EQUATIONS 84

CHAPTER 20: QUADRATIC LINEAR PAIRS 90

CHAPTER 21: RIGHT TRIANGLES 92

CHAPTER 22: QUADRILATERALS: 4 SIDED POLYGONS 97

CHAPTER 23: PERIMETER AND AREA OF
GEOMETRIC FIGURES 99

CHAPTER 24: STUDY OF SOLIDS 100

CHAPTER 25: AREA AND CIRCUMFERENCE OF CIRCLES 104

CHAPTER 26: PROBABILITY 106

CHAPTER 27: STATISTICS AND REAL WORLD APPLICATIONS 120

CHAPTER 28: ERROR IN MEASUREMENT 128

PROBLEM SOLVING STRATEGIES 130

GLOSSARY 135

INDEX 149

1 – SETS AND SET NOTATION

DEFINITIONS

Universe or Universal Set: The set of all possible elements available to form subsets.
> **Example** The set of real numbers.

Set: A group of specific items within a universe.
> **Example** The set of integers.

Subset: A set whose elements are completely contained in a larger set.
> **Example** The set of even integers is a subset of the set of integers.

Complement of a set: Symbols are A', $\overline{A}$ or A^C. A' contains the elements of the universal set that are not in Set A.
> **Example** If the universe is whole numbers from 2 to 10 inclusive, and Set A = $\{2, 4, 6, 8, 10\}$, then $A' = \{3, 5, 7, 9\}$.

Solution Set: All values of the variable(s) that satisfy an equation, inequality, system of equations, or system of inequalities.
> **Example** In the set of real numbers, R, the solution set for the equation $x^2 = 9$ is $\{3, -3\}$ because 3 and -3 are solutions to that particular equation. Write the solution set: SS = $\{-3, 3\}$

SYMBOLS

$\in$ means "is an element of".   **Example** $100 \in$ set of perfect squares.

$\notin$ means "is not an element of".   **Example** $3 \notin$ set of perfect squares.

$\cup$ means the union of two or more sets. The union of two sets contains all the elements in either set.
> **Example** If Set $A = \{2, 4, 6, 8, 10\}$ and Set $B = \{1, 2, 3, 4, 5, 6\}$, then the union of sets A and B, written as $A \cup B$ is $\{1, 2, 3, 4, 5, 6, 8, 10\}$.
> If Set $C = \{12, 18\}$, then $A \cup B \cup C = \{1, 2, 3, 4, 5, 6, 8, 10, 12, 18\}$

$\cap$ means the intersection of 2 or more sets. The intersection of 2 sets contains only the elements that are shared by all sets.

Example Set $A = \{2, 4, 6, 8, 10\}$ and Set $B = \{1, 2, 3, 4, 5, 6\}$, the intersection of sets A and B is: $A \cap B = \{2, 4, 6\}$. If Set $C = \{3, 6, 9\}$ then $A \cap B \cap C = \{6\}$ as it is the only element in all 3 sets.

$\varnothing$ or $\{\ \}$ are symbols for the "empty set" or the "null set." The empty set or null set has no elements in it.

Example If P is the set of negative numbers that are perfect squares of real numbers, then $P = \{\ \}$ or $P = \varnothing$.

- Do not use $\{\ \}$ and $\varnothing$ together. $\{\varnothing\}$ means the set containing the element $\varnothing$. It does not mean the empty set or null set.

Notation: ways to write the elements of a set

Roster Form: A list of the elements in a set (Set A). Ex: $A = \{2, 4, 6, 8, 10\}$

Set Builder Notation: A descriptive way to indicate the elements of a set.

Example The set of real numbers between 0 and 10, inclusive, would be written: $\{x : x \in R, 0 \le x \le 10\}$ or $\{x \mid x \in R, 0 \le x \le 10\}$. The colon or the line both mean "such that." This is read, "The set of all values of x such that x is a real number and 0 is less than or equal to x and x is less than or equal to 10."

Interval Notation: Use $(\,,)$ or $[\,,]$ to indicate the end elements in a list of elements. (and) shows the number is not included, [or] shows that it is.

Example Set $E = \{x: 3 \le x < 6\}$ would be written $[3, 6)$.

Two common symbols for sets of numbers:

R – the set of real numbers. $R = \{x: -\ \}$ The "reals" contain all the numbers from negative infinity to positive infinity.

W – the set of whole numbers. $W = \{0, 1, 2, 3, ...\}$ The whole numbers are 0, 1, 2, 3, and so on. The three dots indicate "and so on" following the pattern established by at least 3 previous elements.

- The 3 dots can also be used on the negative side of a pattern.

Example Integers $< 0 = \{...-3, -2, -1\}$

---

2 – NUMBERS AND ALGEBRA

Sets of Numbers: Groups of numbers. Set notation is used to show the members of a set or to describe the members of the set.

Example Whole Numbers or $\mathcal{W}$ = {0, 1, 2, 3,...}.
This is read, "The set of whole numbers contains the numbers 0, 1, 2, 3, and so on." The "and so on" means that the pattern of numbers shown is to be continued.

Positive Numbers = {$x \mid x > 0$}.
This is read, "The set of positive numbers contains all the values of x such that x is greater than zero." In other words, all the numbers in this set are greater than zero.

Number Line: A representation of all the real numbers in our number system. Each point on a number line corresponds with a REAL NUMBER. Numbers are shown on the number line with zero in the center, the negative numbers on the left side and the positive numbers on the right side of zero. The line extends infinitely in both directions. We often use only a part of the number line — just showing the numbers that are relevant to our current work. Equally spaced tick marks show the placement of the whole numbers and are labeled. All fractions, decimals, square roots, both positive and negative, are located in relation to the labels shown for the whole numbers.

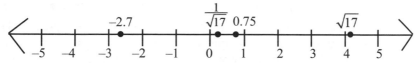

Examples

❶ $\frac{3}{4}$ or 0.75 is located between 0 and 1.

❷ −2.7 is located between −2 and −3.

❸ $\sqrt{17} \approx 4.123$ which is between 4 and 5.

❹ $\frac{1}{\sqrt{17}} \approx 0.243$ 0.243 is between 0 and 1, quite close to zero.

Order: On a number line, the numbers get larger as we read toward the right. In the example on the previous page, –2.7, the number furthest to the left, is the smallest (least) number used. The point for $\frac{1}{\sqrt{17}}$, 0.243, is the next point as we read to the right. It is greater than –2.7, but smaller than 0.75, which is the next number reading to the right. $\sqrt{17}$, or 4.12, is further right than 0.75 which means it is larger than 0.75. $\sqrt{17}$ is the largest number in our example and its point is furthest to the right on the number line. Using < and > signs, we would say: $-2.7 < \frac{1}{\sqrt{17}} < 0.75 < \sqrt{17}$. If you are uncertain where a number is located on a number line, convert the number to a decimal and it will be easier to find on the number line. (Use at least 3 significant digits when changing into decimals.) Then the numbers can be easily compared by examining their relative positions.

Reciprocal: The reciprocal of a number can be formed by making the number into a fraction and then inverting the fraction (flipping it upside down).

Examples of reciprocals *2 and $\frac{1}{2}$, or $\frac{2}{3}$ and $\frac{3}{2}$, or –5 and $-\frac{1}{5}$*

Note: If variables (letters) are used instead of numbers in a general question about the order of numbers, choose a number (like 5) and substitute it into the problem. You can see how that particular number works and then you can usually apply that information to the problem in general.

Example For the set of positive real numbers, is $1 \le x \le \frac{1}{\sqrt{x}}$ true or false? Substitute a number for x that makes sense here, like 9. At first it seems true since $1 \le 9$ --- but 9 is not $\le \frac{1}{3}$ so that makes the expression false if x is 9. One counter-example (See page 9) is enough to make it false, but check once more. Substitute a small number, like 0.04, See if it is still false. Since 1 is not ≤ 0.04, this example is already false. Trying a couple of different numbers and putting them on the number line is helpful if you are "stuck".

Remember a few rules:
    Real Numbers: All the sets below are part of the REAL NUMBER System. The REAL NUMBERS consist of *all* the numbers on a number line. If it were possible to place a point on the number line for each real number, the line would be completely filled in and extend forever (infinitely) in the negative and in the positive directions. (These are also referred to as negative infinity and positive infinity). Real Numbers cannot be listed as a set. Write SS = {Real Numbers} if needed. The symbol (R) also represents the set of real numbers. The Real Numbers are composed of two major sets of numbers called RATIONAL numbers and IRRATIONAL numbers. Each of these sets contains several subsets.

Integrated Algebra Made Easy

1. Rational Numbers: Symbol is (Q). Rational numbers are defined as the ratio of two integers and can be written in the form $\frac{a}{b}$ where a and b are integers and $b \neq 0$ (Division by zero is undefined). It is acceptable for b to be 1, so a rational number like 3 could be written $\frac{3}{1}$. Rational numbers include these sub-sets: (The commonly used symbols and the appropriate elements are shown in set formation).

a) <u>Whole Numbers:</u> $\mathcal{W} = \{0, 1, 2, 3, ...\}$

b) <u>Natural Numbers or Counting Numbers:</u> $\mathcal{N} = \{1, 2, 3, ...\}$

c) <u>Integers:</u> $Z = \{..., -2, -1, 0, 1, 2, ...\}$ *Subsets of integers include*

- **Odd Integers:** Positive or negative. Have a remainder when divided by 2. Odd Integers = $\{..., -5, -3, -1, 1, 3, 5, ...\}$

- **Even Integers:** Positive or negative. Divisible by 2. Include 0. Even Integers = $\{..., -6, -4, -2, 0, 2, 4, 6, ...\}$

- **Positive Integers:** $\{1, 2, 3, ...\}$ [does *not* include 0]

- **Non-Negative Integers** = $\{0, 1, 2, 3, ...\}$ [includes Positives & 0]

- **Negative Integers:** $\{..., -3, -2, -1\}$ [does *not* include 0]

- **Non-Positive Integers:** $\{..., -3, -2, -1, 0\}$ [includes Negatives & 0]

Zero is an even integer, but it has no sign.*

d) <u>Fractions:</u> A ratio of two integers with a denominator that is not equal to zero, like $\frac{2}{3}$ *or* $\frac{5}{4}$.

e) <u>Terminating decimals:</u> Such as 0.24

f) <u>Repeating decimals:</u> Such as $0.\overline{37}$

g) <u>Roots of "perfect" numbers</u> which are rational.

The square root of 25, $\sqrt{25}$,

which is 5, the cube root of -8, $\sqrt[3]{-8}$,

which is $-2\sqrt{0.04}$ $= 0.2$, or $\sqrt{225} = 15$.

<u>Note</u>: There is an infinite number of rational numbers and they cannot be written as a listed set. Write SS = {rational numbers} if needed.

---

2. Irrational Numbers: Whole numbers which cannot be expressed as the ratio of two integers.

a) **Non-repeating**, **nonterminating decimals** like 0.354278...

b) The **roots of numbers which do not calculate to a rational number**. The square root of 12 which ≈ 3.4641016.

c) **Pi *or* π** is also irrational. The "π" *button on the calculator must be used to compute formulas that contain* π. Rounding to 3.14 is *not* permitted.

Note: When a calculator is used to work with irrational numbers, rounding is not permitted. To "show all work" write down on the paper all the digits in the calculator display. Leave the entire number in the calculator until the end of the problem. When directions say to round, it should be the final step of a problem.

The diagram below represents the relationship between subsets of real numbers. The universal set, shown by the rectangular box, is the Real Numbers. Subsets are shown as circles. Subsets of the Reals are Irrational numbers and Rational numbers. Subsets of Rationals are Integers and its subsets are Whole numbers and Counting numbers.

Real Numbers

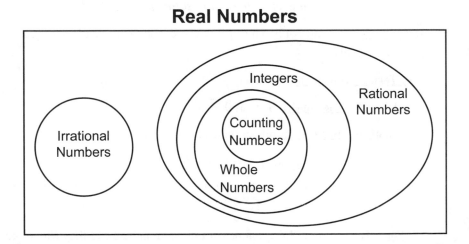

---

3 – PROPERTIES AND LAWS

WHY DO WE CARE ABOUT THE PROPERTIES? These fundamental laws allow us to manipulate the terms in an equation. We can move things around, transfer parts of an equation to the opposite side of the equal sign, and use the identities to help isolate the variable. Closure applies to polynomials. The addition of two polynomials results in a sum that is a polynomial. The associative property is also applicable to polynomials under addition. This means the grouping can be changed. The additive identity of zero exists and each polynomial has an additive inverse. The commutative property is also true for the addition of polynomials. Combining these properties and identities gives us the flexibility we need to solve problems and equations involving polynomials algebraically.

These laws and properties apply to real numbers and algebraic equations.

<u>**Commutative Property of addition and multiplication**</u>: The POSITION of the numbers can be changed without changing the answer.

Addition: $a + b = b + a$

Example $5 + 1 = 1 + 5$

Multiplication: $ab = ba$

Example $(-5)(-2) = (-2)(-5)$

Note: The commutative property is *not true* with respect to subtraction or division of real numbers.

Subtraction: $5 - 3 \neq 3 - 5$

Division: $\dfrac{2}{3} \neq \dfrac{3}{2}$

<u>**Associative Property of addition and multiplication**</u>: The GROUPING of the numbers can be changed without changing the answer.

Addition: $(a + b) + c = a + (b + c)$

Example $(5 + 4) + 2 = 5 + (4 + 2)$

Multiplication: $(ab)\,c = a\,(bc)$

Example $(3 \cdot 2)(4) = 3(2 \cdot 4)$

Note: The associative property is *not true* for subtraction or division of real numbers.

Subtraction: $(3 - 5) - 12 \neq 3 - (5 - 12)$

Division: $(12 \div 3) \div 4 \neq 12 \div (3 \div 4)$

<u>**Distributive Property of multiplication over addition**</u>: Multiply each term on the inside by the number or letter on the outside of the parenthesis.

Examples ❶ $a(b + c) = ab + ac$

❷ $2(3 + 5) = 2(3) + 2(5)$

❸ $-3(x + 2) = -3x - 6$

❹ $-5(2x + 5 + 2x) = -5(4x + 5) \Rightarrow$ then $-20x - 25$ is the final answer.

❺ $2x(3x - 4) = 6x^2 - 8x$

Use the positive or negative sign before the outside number with the number when you multiply. This removes the parenthesis and takes care of the needed sign changes. Combine terms inside the parenthesis before you use the distributive property if possible. If the number on the outside is a fraction, be sure to multiply each term inside the parenthesis by the entire fraction.

SPECIAL NOTE: **Use the Distributive Property for Subtraction Problems**

A parenthesis with subtraction sign in front of it can be handled using the Distributive Property. Put "1" outside the parenthesis between "–" and the ( ). Use the Distributive Property by multiplying by –1. This will automatically take care of the sign changes needed for subtraction and remove the ( ), then simplify (Combine like terms).

Examples

❶ $3x - 4 - (7 + x) \Rightarrow 3x - 4 - 1(7 + x) \Rightarrow 3x - 4 - 7 - x \Rightarrow 2x - 11$

❷ If a problem says "take $(3x - 4)$ from $(12x + 5)$", then do this:

Steps: 1) Set-up $(12x + 5) - (3x - 4)$ "From" number is first in set-up

     2) Get ready $(12x + 5) - 1(3x - 4)$ Put a "**1**" between – and $(3x - 4)$.

     3) Remove ( ) $12x + 5 - 3x + 4$ Use Distributive Property and multiply by –1.

     4) Simplify $9x + 9$ Collect Like Terms.

❸ If the directions say "SUBTRACT" and the problem is already set up:

Step 1) is already done.	**Step 2)**	**Step 3), then 4)**
$(5x^2 - 12x + 14)$	$(5x^2 - 12x + 14)$	$5x^2 - 12x + 14$
$- (x^2 + 13x - 25)$	$-1(x^2 + 13x - 25)$	$+ (-x^2 - 13x + 25)$
		$4x^2 - 25x + 39$

<u>*Note*</u>: Sometimes the *directions* say "subtract these problems" and there may *not* be a subtraction sign in front of the lower expression. Put in your own ( ) and –1 if needed, then go on. READ the directions.

Integrated Algebra Made Easy

Identity Element: An element that can be used with the given operation on any member of the set without changing its value is called the identity element for that operation. If * is a binary operation on set **S**, then "*e*" is an identity element if, for all *a* in **S**, *a* * *e* = *e* * *a* = *a*.

REAL NUMBER IDENTITIES

Additive Identity: Zero (0) is the additive identity in the real number system. A number is unchanged by adding zero to it or by adding the number to zero.

$$5 + 0 = 0 + 5 = 5$$

Multiplicative Identity: One (1) is the identity element for multiplication in the real number system. A number multiplied by one is unchanged.

$$(5)(1) = (1)(5) = 5.$$

Inverse Element: In a system with an identity element "*e*" there is an inverse "*b*"such that *a* * *b* = *b* * *a* = *e*. When the given operation is performed on an element and its inverse, the result is the identity element.

REAL NUMBER INVERSES

Additive Inverse: The Additive Inverse of a number in the real number system is the opposite number. (The same number but with the opposite sign.) Since 0 is the identity element for addition in the real numbers, then –4 is the additive inverse of 4.

$$(-4) + (4) = 0$$
Since $(-3.2) + (3.2) = 0$ and $(3.2) + (-3.2) = 0$,
3.2 is the additive inverse of –3.2.

Multiplicative Inverse: The reciprocal of a real number is its multiplicative inverse. The product of a number and its reciprocal is 1 which is the identity element for multiplication of real numbers.

$$-3\left(-\frac{1}{3}\right) = 1 \text{ and } \left(\frac{2}{3}\right)\left(\frac{3}{2}\right) = 1$$

The multiplicative inverse of a negative number is still a negative number — be sure and keep the reciprocal as a negative number.

Note: The reciprocal of 1 is 1 and the reciprocal of –1 is –1. These properties, laws, and definitions can be applied to other systems of mathematics that may include letters and operations that we are not familiar with.

Multiplication Property of Zero: Any number times zero = 0.

---

4 – MATHEMATICAL SYSTEMS

Operation: A rule or process that is performed on the elements of a set. It may be familiar like +, –, ×, or ÷. The operation may also be shown as a symbol like @ or * or ⊗. The symbol may represent an operation or it may represent the use of a formula.

Binary Operation: An operation which uses exactly two elements of a set and results in an element which may or may not be an element of that set.

Examples

① Addition on the set of real numbers: $3 + 4 = 7$

② $A * B = C$ using * as the operation on the set of letters of the alphabet. (We don't know exactly what * did to A and B, but we do know that after * was performed on two elements of the set given, the result was C.)

Mathematical Systems: A Math System can consist of *sets of elements which are numbers* or *sets of elements that are not numbers* but which can be used in a mathematical structure. The set of real numbers is the set used unless some other set is specified.

Often a chart is used that shows all the possible elements as shown here. The elements are shown in a vertical column and a horizontal row, using the same order. The top left corner position is where the operation is shown. The inside columns and rows of the chart are filled with the results of the operation on each pair of elements given. The chart can be used to evaluate the result of the given operation on various elements in the system.

• The example given has elements that are not numbers.
• However real numbers can be used with the 4 operations (+, –, ×, or ÷).

Example Using the given system, show: **E @ U**

Steps:

1) Locate the first element used in the problem, **E**, in the left column of the chart. Imagine (or draw) a horizontal line across the chart from there.

@	T	U	N	E
T	T	U	N	E
U	U	E	T	N
N	N	T	E	U
E	E	N	U	T

2) Locate the second element, **U**, in the top row of the chart. Draw or imagine a vertical line down through the letters of that column.

3) The intersection of the horizontal and vertical lines is the answer. The lines intersect at **N** so **E @ U = N**. Always start in the left column using the first element in the problem.

@	T	U	N	E
T	T	U	N	E
U	U	E	T	N
N	N	T	E	U
E	E	N	U	T

Integrated Algebra Made Easy

For compound operations, do the work inside the parentheses, then work from left to right.

Example Answer the questions using figure: (N @ U) @ E = T @ E = E

1 • Does this system have closure and why?
 • To solve, test each pair of elements to see if the answer is always within the system. **T @ T = T; N @ E = U**, etc.
 • Since the result is always in the system, it has closure.

2 • Is the system commutative under @ and why?
 • Again, test each pair of elements for commutativity.
   T @ N = N and N @ T = N; E @ N = U and N @ E = U, etc.
 • This is true of any two elements in the set so it is commutative.

3 • Is the system associative under @ and why?
 • **(U @ T) @ E = N and U @ (T @ E) = N.**
 • This is maintained throughout the elements if tested, therefore the system is associative.

4 • Is there an identity element?
 • What is it?
 • **T @ T = T, U @ T = U, N @ T = N, and E @ T = E** so **T** is the identity element.

5 • What is the inverse of **T**? of **U**? of **N**? of **E**?
 • Remember that using an element with its inverse equals the identity element.
 • **T** and **T** are inverses as are **U** and **N**, **N** and **U**, and **E** and **E**.

PROPERTIES OF MATH SYSTEMS:

Closure: A set, **S**, is closed under a binary operation (*) if for all "*a*" and "*b*" in **S**, *a* * *b* equals an element of set **S**. In other words, if a binary operation performed on any two members of a set always results in a member of the set, then the set has closure with respect to that operation. To determine if a set has closure with respect to an operation, perform the operation using all members of the set. If the answer is always a member of the set, it has closure with respect to the operation.

Examples of closure on the set of whole numbers, $\mathcal{W}$:

1 **Addition:** $2 + 5 = 7$. Two and five are elements of $\mathcal{W}$. Seven is also an element of $\mathcal{W}$. There are no examples of addition of whole numbers which would not result in a whole number answer. The set has closure or we say the system "is closed" with respect to addition.

❷ **Multiplication:** $2 \cdot 6 = 12$. Here too, 6, 2, and the result of $2 \cdot 6$, 12, are all members of the set of whole numbers. There are no whole numbers that can be multiplied together that will give an answer that isn't a whole number. W has closure with respect to multiplication.

> *Note*: Subtraction and Division: Whole numbers are not closed with respect to subtraction or division.

Commutative Property: An operation (*) is commutative on a set **S** if for all a and b in **S**, $a * b = b * a$. In other words, if the position of the two elements can be reversed and give the same answer, the set is commutative under that operation. (See also page 5)

Associative Property: An operation is associative on a set, **S**, if for all a, b, and c, in **S**, $(a * b) * c = a * (b * c)$. In other words, in sets that are associative under a particular operation, it doesn't matter what grouping is used on the elements, the answer will be the same. (See also page 5)

Distributive Property: The Distributive Property allows two different operations to be handled in the same problem. In the set of real numbers, multiplication distributes over addition and subtraction. The distributive property is essential in solving algebra problems. In a problem with two operations, @ and *, @ is said to distribute over * if, for all values of a, b, and c, the following is true: $a @ (b * c) = (a @ b) * (a @ c)$. (See also page 6)

Counter-Example: An example which gives an answer that shows a proposal is *not* true in at least one case. In trying to prove closure with W, if an example can be found which gives a result that is not in the set of whole numbers, then the set is not closed with respect to that operation.

While $12 - 7 = 5$ seems to demonstrate closure of whole numbers under subtraction, the counter-example $3 - 7 = -4$ contradicts it. (–4) is not a whole number. Whole numbers are not closed with respect to division either. While $20 \div 4 = 5$ seems to prove closure, $10 \div 4 = 2.5$ disproves it because 2.5 is not a member of the set of whole numbers.

---

5 – SIGNED NUMBERS

<u>Positive</u> (+): numbers greater than zero. (*NON-POSITIVE* means negatives and zero.)

<u>Negative</u> (–): numbers less than zero. (*NON-NEGATIVE* means positives and zero.)

* ZERO HAS NO SIGN - it is neither positive nor negative.

<u>Absolute Value</u> (| |): Absolute value is the distance from the number inside the symbol (| |) to zero on a number line.

- It is always positive.
- Do the inside work first, then make that answer positive.
- The absolute value of 0 is zero.
- A negative sign to the left of the absolute value sign means to find the absolute value of the "inside" work and then make that answer negative. [See Ex. 3]
- A number in front of the absolute value sign means to find the absolute value inside the symbols and then multiply by the outside number. [See Ex. 4]

Examples ❶ $|5| = 5$

❷ $|-8| = 8$

❸ $-|3 + 4| = -|7| = -7$

❹ $2|2 - 15| = 2|-13| = 2(13) = 26$

<u>Operations with signed numbers</u>: Work left to right using the correct order of operations. If there are more than 2 numbers to work on, do them in pairs -- always using the order of operations. (See page 13)

<u>Addition</u>: LIKE signs (all + or all –; all the same): Keep the sign, add the absolute values.

Examples ❶ $12 + 5 = 17$

❷ $(-5) + (-3) = -8$

❸ $(-4) + (-2) + (-7) = (-6) + (-7) = -13$

UNLIKE signs (one +, one –; mixed or different signs) Find the difference between the absolute values of the numbers. Keep the sign of the number with the larger absolute value.

Examples   ❶ $12 + (-8) = 4$

❷ $20 + (-30) = -10$

❸ $(-8) + 5 + (-9) = (-3) + (-9) = -12$

<u>Subtraction</u>: Change subtraction sign to an addition sign, and change the sign of the number <u>after</u> the subtraction sign. Now follow the addition rules.

Examples   ❶ $(-9) - (+6) = -9 + (-6) = -15$

❷ $12 - 8 + 15 = 12 + (-8) + 15 = 4 + 15 = 19$

❸ $(-10) - (-7) - 6 = (-10) + 7 + (-6) = -9$

Shortcut: Consider any of the above problems as an addition problem.

- Process the subtraction sign if it is in front of a parenthesis by changing the sign of whatever is in the parenthesis. (see also page 8)

- Next, "collect" all the positive numbers and find their sum.

- "Collect" all the negative numbers and find their sum.

- Then find sum of the positive and negative numbers using the rules of addition (find the difference of the numbers and use the sign of the number with the higher absolute value).

Example of shortcut
Steps:

1)	Original	$-5 + 8 - (-6) + 4 - 28 =$
2)	Process $-( )$	$-5 + 8 + 6 + 4 - 28 =$
3)	Sum + #'s and – #'s	$(8 + 6 + 4) + (-5 + -28) =$
4)	Add	$18 + (-33) =$
5)	Answer	-15

Multiplication and Division (2 numbers):

Like Signs: (2 negatives or 2 positives). Answer is positive.

Examples ❶ $(3)(4) = 12$

❷ $(-5)(-4) = 20$

Unlike Signs: (one of each). Answer is negative.

Example $(-2)(3) = -6$

(When doing multiplication of *more than 2 signed* numbers, do the numbers in pairs.)

Example $(-2)(4)(-3) = (-8)(-3) = +24;$

Think: (-2) times $(+4)$ is (-8), then (-8) times (-3) is $(+24)$.

Division in Fractions: The fraction line acts like a grouping symbol. Using the correct order of operations, do all the work in the numerator (top) of the fraction. Do all the work in the denominator (bottom) of the fraction using the correct order of operations. Then divide the top by the bottom. Follow the rules for signs as needed in each part of the problem.

Example $\dfrac{(14-2)-(3+3)}{-12} = \dfrac{12-6}{-12} = \dfrac{6}{-12} = -\dfrac{1}{2}$ or -0.5

Division by Zero is undefined: If a fraction has zero as a denominator, the division cannot be performed. The fraction is said to be "undefined."

Order of Operations

When performing operations on numbers (or on algebraic terms), it is necessary to follow these rules that tell which part of the work to do first, second, etc. Read the problem from left to right, just as you read words, and follow these steps:

1) Parenthesis — Do whatever work you can inside a parenthesis to simplify the problem. The fraction line in a fraction acts like a parenthesis -- do the work in the numerator of the fraction and the work in the denominator of the fraction before continuing.

2) Exponents — Apply any exponents in the problem to their bases. This includes a parenthesis with an exponent or a term with an exponent raised to another power.

3) Multiply and Divide — Do these in order as they appear as you read left to right. This includes using the distributive property.

4) Add and Subtract — Do these in order as they appear reading left to right.

Example $-3(4+1)^2 + 4 \div 2 + \dfrac{3+3}{2}$

Steps:

1) Parenthesis $(4+1) = (5)$

     Grouping symbols(Fraction bar) $3 + 3 = 6$ $-3(5)^2 + 4 \div 2 + \dfrac{6}{2}$

2) Exponent $(5)^2 = (5)(5) = 25$ $-3(25) + 4 \div 2 + \dfrac{6}{2}$

3) Multiply $-3(25) = -75$ $-75 + 4 \div 2 + \dfrac{6}{2}$

4) Divide $4 \div 2 = 2$

     $6 \div 2 = 3$ (simplify the fraction) $-75 + 2 + 3$

5) Collect like terms -- add the numbers $-73 + 3 = -70$
     using the appropriate rules for addition
     of signed numbers.   Add/Subtract

Note: Remember to always work left to right -- within a parenthesis or in a
       fraction as well as across the entire problem.

The example below shows a combination problem that requires several steps to
simplify. (See also Monomials and Polynomials page 17)

Example $3(x + 3x - 1)^2 - (8x^2 + 4)$

Steps:

1) Do work inside the ( ) first. $3(4x - 1)^2 - (8x^2 + 4)$

2) Exponent applied to$(4x - 1)$ only. $3(4x - 1)(4x - 1) - (8x^2 + 4)$

3) Use F. O. I. L. $3(16x^2 - 4x - 4x + 1) - (8x^2 + 4)$

4) Collect like terms inside the ( ). $3(16x^2 - 8x + 1) - (8x^2 + 4)$

5) Use the Distributive Property
     to multiply and remove ( ). $48x^2 - 24x + 3 - 8x^2 - 4$

6) Collect like terms to simplify. $40x^2 - 24x - 1$

---

6 – MONOMIALS AND POLYNOMIALS

Term: It indicates a product and may have one factor or many factors. It may contain numerical factor called a coefficient, one or more variables which may have exponents. A coefficient or an exponent of "1" is not written.

> **Example** x, $3x^2$ *or* xy

Coefficient: A numerical factor written to the left of a variable. A coefficient of one is understood and is not written.

> **Example** $6x^2$ the coefficient is 6. *or* y^3 the coefficient is 1.

Constant: A coefficient without a base and exponent. It's just a number.

Like Terms: Have the same LETTER BASE and the SAME EXPONENT. They can be combined by adding the numerical coefficients (with their signs) and keeping the base (variable) and its exponent UNCHANGED.

> **Example** $5x^3 + 4x^3 = 9x^3$ *or* $x^2 - 6x^2 = -5x^2$

Unlike Terms: Have different variables and/or different exponents. These *cannot* be collected or *added*. (Unlike terms can be multiplied when indicated.)

> **Example** The following cannot be simplified; $3x + 4$, $x^2 + 4x$, *or* $x^2 + y^2$

Expression: Two or more terms, with or without a constant, that are connected with + or −.

> **Example** $x + y$, $2x^2 + 5$, $6xy + 7$, $3x^2 + 2x + 4$

Polynomial: Terms are separated by + or − to form polynomials. Monomials, binomials, and trinomials are polynomials with special characteristics.

1) **Monomial:** A *single algebraic term* or the *product* of algebraic terms.

> **Example** $4x^2$ *or* $12xy$ *or* $8x^2yz^2$ *but not* $\dfrac{3x}{2y}$.

2) **Binomial:** The sum or difference of 2 (unlike) monomials that cannot be combined.

> **Example** $4x^2 + 12x^2y$ *or* $5m - 2$ *or* $x^2 - 4$

3) **Trinomial:** The sum or difference of 3 monomials that cannot be simplified or combined.

> **Example** $3x + 4y + 8$ *or* $x^2 + 4x + 5$

Equation: An expression that is equal to a constant or to another expression and is connected using an = sign.

> **Example** $x = 5$, $x + 3 = 2y$, $x^2 - 6x - 7 = 0$

OPERATIONS WITH POLYNOMIALS

Addition: Remove any parenthesis using the distributive property if needed. (You can always put a "1" in front of a ( ) if it is easier for you.) Then combine like terms.

Subtraction: Multiply the terms in the polynomial to be subtracted by –1 (use the distributive property). This removes the parenthesis and takes care of the SIGN CHANGES. Then ADD by combining like terms.

Example $(3x - 2) + (5x - y) - (2x - 4)$

Steps:
1) Put in "1's": $1(3x - 2) + 1(5x - y) - 1(2x - 4)$
2) Use the distributive property: $3x - 2 + 5x - y - 2x + 4$
3) Simplify or collect like terms: $6x - y + 2$

Example From the sum of $(x + 2y)$ and $(2x - y)$, subtract $(3x - y)$.

Steps:
1) Put brackets around the sum first: $[(x + 2y) + (2x - y)] - (3x - y)$
2) Simplify inside the bracket: $[x + 2x + 2y - y] - (3x - y)$
3) Put in "1's": $1(3x + y) - 1(3x - y)$
4) Use the distributive property: $3x + y - 3x + y$
5) Simplify or collect like terms: $2y$

Example Subtract $5x - 2y$ from $12x - 5$

Steps:
1) Put the "from" expression first: $(12x - 5) - (5x - 2y)$
2) Put in "1's": $1(12x - 5) - 1(5x - 2y)$
3) Use the distributive property: $12x - 5 - 5x + 2y$
4) Simplify or collect like terms: $7x + 2y - 5$

*****EXPONENTS DO NOT CHANGE IN ADDITION/SUBTRACTION*****

Multiplication: ANY Two Terms (like or unlike) can be multiplied.
Exponents of like variables are added in multiplication.

Monomial • Monomial: Use this **Example** $(5x^2 y)(-3x y^3 z)$

Steps:
1) Multiply the numerical coefficients $(5)(-3) = -15$
2) Multiply the LIKE LETTER bases by ADDING their EXPONENTS $(x^2)(x) = x^3$ and $(y)(y^3) = y^4$
3) Multiply the unlike letter bases by simply writing them down as part of the product. $-15(x^3)(y^4)(z)$
4) Write the answer as a product. $-15 x^3 y^4 z$

Monomial • Polynomial: Use the distributive property (See page 6)

Example $3x(2x^2 - 6x + 1) = 6x^3 - 18x^2 + 3x$

Integrated Algebra Made Easy

Binomial • Binomial: Each term in the first binomial must be multiplied by each term in the 2nd binomial. Use **F. O. I. L.** The letters in FOIL refer to the position of the terms in each parenthesis. They tell what to multiply. Multiply: First terms together, **O**utside terms together, **I**nside terms together, and **L**ast terms together. Then simplify. Use example below:

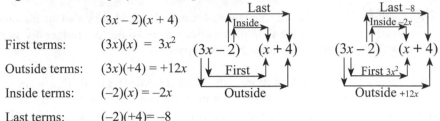

First terms: $(3x)(x) = 3x^2$

Outside terms: $(3x)(+4) = +12x$

Inside terms: $(-2)(x) = -2x$

Last terms: $(-2)(+4) = -8$

Collect like terms and simplify: $3x^2 + 12x - 2x - 8 = 3x^2 + 10x - 8$

Polynomial • Polynomial: Multiply each term in the first parenthesis by each term in the 2nd.

$$(x-2)(x^2 - 3x + 2) = x(x^2 - 3x + 2) - 2(x^2 - 3x + 2) =$$
$$x^3 - 3x^2 + 2x - 2x^2 + 6x - 4 = x^3 - 5x^2 + 8x - 4$$

Division: In division, exponents of like variables are subtracted. SIMPLIFYING fractions or reducing fractions involve these same processes. Remember that division by zero is undefined (It is not possible to divide by zero).

Monomial ÷ Monomial *or* Monomial/Monomial *or* $\dfrac{\text{Monomial}}{\text{Monomial}}$

$-15x^5 y^7 \div 3x^2 y^2$

Steps: 1) Divide the numerical coefficients and write that quotient in simplest form. $(-15) \div (3) = -5$

2) Divide LIKE BASES by SUBTRACTING their EXPONENTS. $x^5 \div x^2 = x^3$ and $y^7 \div y^2 = y^5$

3) Write the answer as a product. $-5x^3 y^5$

Example $14x^2 \div 7y$ (*x and y are Unlike terms*)

Steps: 1) Divide the numerical coefficients: $14 \div 7 = 2$

2) x and y are unlike bases so they cannot be divided. Write them in fraction form $\dfrac{x^2}{y}$

3) Write the result as a product: $2\left(\dfrac{x^2}{y}\right) = \dfrac{2x^2}{y}$

Integrated Algebra Made Easy
19

Polynomial ÷ Monomial: When dividing a polynomial by a monomial, separate the polynomial into the appropriate number of fractions using the monomial denominator for each fraction. Then follow the rules above.

Example $\dfrac{2x^2 - 4}{2} = \dfrac{2x^2}{2} - \dfrac{4}{2} = x^2 - 2$

Example $\dfrac{5x^2 + 4}{x}$ The constant term, 4, cannot be divided by x so the fraction is already considered to be simplified as much as possible. No further division is possible. (See also - Algebraic Fractions page 34)

Division of a Polynomial by a Binomial: There are two ways to divide a polynomial by a binomial. In the first method, the numerator and denominator of the fraction are factored (see page 38) and one of the binomial factors matches the binomial in the denominator. The two matching binomials can be cancelled. The quotient is the remaining factor of the numerator written as a fraction over any remaining numerical factor in the denominator.

Example

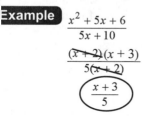

Factor and cancel. Denominator becomes

1 so we don't need to write it.

Answer.

Sometimes the denominator has a Greatest Common Factor (GCF) which needs to be factored out before doing the cancellation. (see page 35.)

Example $\dfrac{x^2 + 5x + 6}{5x + 10}$ Denominator has GCF of 5

$\dfrac{(x + 2)(x + 3)}{5(x + 2)}$ Factor both the numerator & denominator. Cancel. "5" remains in the denominator.

$\boxed{\dfrac{x + 3}{5}}$ Answer.

Long Division: Since not all polynomials can be factored easily, sometimes long division is required. This process is very much like long division with numbers.

(See next page for long division)

Long Division – Use the long division format used in arithmetic. The dividend and the divisor must be written using descending powers of the variable.

Example $(2x^2 - 3x - 20) \div (x - 4)$

Steps:

$$\begin{array}{r} 2x+5 \\ (x-4)\overline{)2x^2-3x-20} \end{array}$$

1) Divide the first term of the dividend $(2x^2)$ by the first term of the divisor (x) to get the first term of the quotient $(2x)$.

$$\begin{array}{r} -(2x^2-8x) \\ \hline +5x-20 \\ -(5x-20) \\ \hline 0 \end{array}$$

2) Multiply the complete divisor $(x - 4)$ by the term in the quotient found in step 1 $(2x)$. Position this product $(2x^2 - 8x)$ to be subtracted from the dividend.

3) Subtract. (Remember to add the opposite). Bring down the next term of the dividend. (-20)

Check:

$(x - 4)(2x + 5) =$

$2x^2 - 3x - 20$

4) Repeat from step 1 using the result from step 3, $(5x - 20)$, as the new dividend.

5) Check by multiplying the quotient by the divisor.

Undefined Algebraic Fraction: When the denominator of a fraction is zero, the fraction is undefined. To find the values of the variable for which a fraction is undefined, make an equation with the denominator = 0. Solve for the variable. The solution(s) for this equation are the values of the variable that make the fraction undefined.

Example $\dfrac{x^2+x-6}{2x+8}$

$2x + 8 = 0$

$2x = -8$

$x = -4$ *Note*: When $x = -4$, the fraction is undefined. Therefore x cannot be -4

7 – EXPONENTS

An exponent indicates how many times to use its base as a factor to make a product. An exponent is used only with whatever number or term is directly to its left. If the exponent is outside a parenthesis, then use the exponent with everything in the parenthesis. If the exponent is next to a term instead of a parenthesis, it goes only with its immediate neighbor to the left. When simplified, a problem with exponents should have an answer where only the variables still have exponents. Numbers with exponents should be multiplied out.

Examples ❶ $(3xy)^3 = (3xy)(3xy)(3xy) = 27x^3y^3$

❷ $(-4)^2 = (-4)(-4) = 16$

❸ $-4^2 = -(4)(4) = -16$

(Check this one out! The exponent goes with the $(2x)$ and not with the -3)

❹ $-3(2x)^3 = -3(2x)(2x)(2x) = -3(8x^3) = -24x^3$

Multiplication: When the bases are alike, ADD the EXPONENTS. MULTIPLY the COEFFICIENTS.

Examples ❶ $(4x\,y^3)(5xy^2) = 20\,x^2y^5$

❷ $(2x^4)(-3x^{-2}) = -6x^2$

Division: When the bases are alike, SUBTRACT the EXPONENTS. DIVIDE the COEFFICIENTS.

Example $\dfrac{6x^3y^4}{2x^2y} = 3x^{3-2}y^{4-1} = 3xy^3$

Powers: When a term with an exponent is used as a base with another exponent, it is called "raising a power to a power." Coefficients are treated as if they have an exponent of 1. Multiply each exponent inside the parenthesis by the exponent on the outside. Be sure to include the "invisible" exponent 1 next to the coefficient. Then simplify if possible.

Examples ❶ $(x^5)^4 = x^{20}$

❷ $(x^2y)^3 = x^6y^3$

❸ $(3x^4y^2)^3 = (3^1x^4y^2)^3 = 3^3x^{12}y^6 = 27x^{12}y^6$

Zero as an exponent: ANY base (except 0) with an exponent of 0 is equal to **1**. Replace the indicated base with 1 in the problem before continuing.

Note: A base of zero raised to the zero power $(0)^0$ is undefined.

Examples ❶ $x^0 = 1$

❷ $x^2 y^0 = x^2 \cdot 1 = x^2$

❸ $4m^0 = 4(1) = 4$

❹ $[42(3250)^3(x^4)]^0 = 1$

Negative Exponents: Make a fraction putting any base with a negative exponent in the denominator. This will make the exponent positive. The coefficient and any variable with positive exponents are the numerator. (Negative exponents are undefined if the base = 0.)

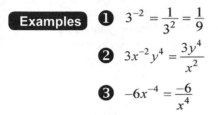

Examples ❶ $3^{-2} = \dfrac{1}{3^2} = \dfrac{1}{9}$

❷ $3x^{-2}y^4 = \dfrac{3y^4}{x^2}$

❸ $-6x^{-4} = \dfrac{-6}{x^4}$

Degree of Polynomials: The degree of a polynomial in one variable is the highest power of the variable.

Examples ❶ $x + y$ is a first degree polynomial. (Also called linear.)

❷ $3x^2 + 2x + 5$ is a 2nd degree polynomial. (Also called quadratic.)

❸ $x^3 + 4$ is a 3rd degree polynomial. (Also called cubic.)

Standard Form: Put the terms in order so that the exponents of one variable are descending (highest exponent is first, then next lower, etc.). Remember: A constant term without a variable must be included, it is really showing a variable with the exponent of zero which makes the variable equal to 1.

Example This is a 5th degree polynomial put in standard form.

$$3x^4 - 4x^2 + 12x^5 + 4 = 12x^5 + 3x^4 - 4x^2 + 4$$

8 – SCIENTIFIC NOTATION

Scientific notation is used to enable people using very large or very tiny numbers in their work to write the number using a "shortcut." It uses powers of 10 which is the base of our number system as a multiplier of manageable size factors written in a standard way. The standard form of the factor is a number greater than or equal to one and less than ten. Generally two decimal places are shown. Here is the method:

Steps:

1) Move the decimal point in the number to the position immediately following the first significant digit. Count the number of places the decimal is moved. (Remember that if a decimal isn't visible, it is at the end of the number.)

2) Show that the new number must be multiplied by a power of 10 with an exponent matching the number of places you had to move the decimal point.

3) Power of ten - Exponents: If the original number is greater than or equal to 10, the exponent for ten will be positive. If the original number is less than 1, the exponent is negative. (If the number was at least 1 and less than 10 to start with, the exponent is 0. It would be unusual to write a number in scientific notation with a zero exponent.)

Examples

❶ $5,830,000. = 5.83 \times 10^6$ *or* $(5.83)(10^6)$

❷ $0.000261 = 2.61 \times 10^{-4}$ *or* $(2.61)(10^{-4})$

❸

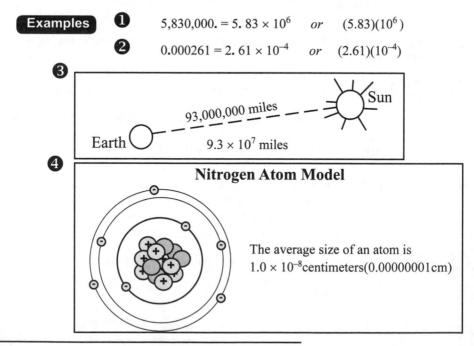

❹

Nitrogen Atom Model

The average size of an atom is 1.0×10^{-8} centimeters(0.00000001cm)

Evaluating problems using powers of 10 (scientific notation).

1. Translate the numbers into scientific notation.
2. Use the commutative property to rearrange the numbers so the coefficients are together and the factors of 10 are together.
3. Do the appropriate operation to the coefficients and use the laws of exponents to process the powers of 10.
4. Translate back into expanded numerical form if required.

Examples

❶ $(20000)(0.00000008) = (2.0)(10^4)(8.0)(10^{-8})$

$$= (2.0)(8.0)(10^4)(10^{-8})$$

$$= (16)(10^{4+(-8)})$$

$$= (16)(10^{-4})$$

$$= 0.0016$$

❷ $\dfrac{650,000}{.005} = \dfrac{(6.5)(10^5)}{(5.0)(10^{-3})}$

$$= (1.3)(10^{5-(-3)}) = (1.3)(10^8)$$

$$= 130,000,000$$

❸ $\dfrac{(4000)(0.00002)}{(200,000)\,(0.0008)} = \dfrac{(4.0)(10^3)(2.0)(10^{-5})}{(2.0)(10^5)(8.0)(10^{-4})}$

$$= \dfrac{(4.0)(2.0)(10^3)(10^{-5})}{(2.0)(8.0)(10^5)(10^{-4})}$$

$$= \dfrac{(8.0)(10^{-2})}{(16.0)(10)}$$

$$= (0.5)(10^{-2-1}) = (0.5)(10^{-3})$$

which is equivalent to $(5.0)(10^{-4})$

$$= 0.0005$$

9 – EVALUATING EXPRESSIONS & FORMULAS

Evaluate – means to find the numerical value of a mathematical statement that contains variables. (Using more student friendly words — find the number that is the answer.) In order to evaluate an expression containing variables or a formula, we must have the numerical values of the variables. Substitute the numerical value of the variable in place of the variable and do the math. Make sure to use the order of operations correctly. Use parentheses to show the substitution so you can avoid errors with positive and negative numbers. Do not round the answer unless directed to do so.

> _Note_: An algebraic expression does not usually use an "=" sign. However, to show the progress of the work as substitution and subsequent operations are performed, = is used to make the work easy to follow.

Evaluate a _formula_: Find the number of degrees Celsius if the temperature is currently 68 degrees Fahrenheit. Use the formula below:

Example

$$C = \frac{5}{9}(F - 32)$$

Substitute (70) for F $\quad C = \frac{5}{9}[68 - 32]$

$$C = \frac{5}{9}(36)$$

$$C = 20°$$

Evaluate an _algebraic expression_: Evaluate each of these expressions
if $x = 5$ and $y = -6$. Remember, the = just shows the sequence of the work.

Examples

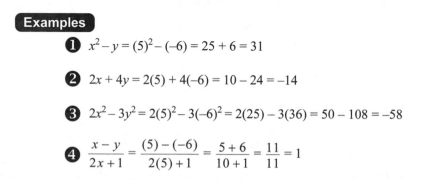

❶ $x^2 - y = (5)^2 - (-6) = 25 + 6 = 31$

❷ $2x + 4y = 2(5) + 4(-6) = 10 - 24 = -14$

❸ $2x^2 - 3y^2 = 2(5)^2 - 3(-6)^2 = 2(25) - 3(36) = 50 - 108 = -58$

❹ $\dfrac{x - y}{2x + 1} = \dfrac{(5) - (-6)}{2(5) + 1} = \dfrac{5 + 6}{10 + 1} = \dfrac{11}{11} = 1$

Evaluate an _absolute value_ expression: First perform any operations that are contained inside the absolute value symbol. Next, take the absolute value of that result. Use that positive number with any other operations required in the expression to complete the evaluation.

Examples

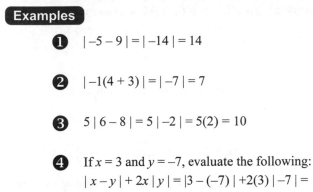

❶ $|-5-9| = |-14| = 14$

❷ $|-1(4+3)| = |-7| = 7$

❸ $5|6-8| = 5|-2| = 5(2) = 10$

❹ If $x = 3$ and $y = -7$, evaluate the following:
$|x - y| + 2x|y| = |3 - (-7)| + 2(3)|-7| =$
$|10| + 6|-7| = 10 + 6(7) = 10 + 42 = 52$

Evaluate an _exponential expression_:

Examples

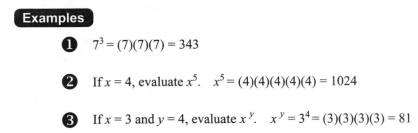

❶ $7^3 = (7)(7)(7) = 343$

❷ If $x = 4$, evaluate x^5. $x^5 = (4)(4)(4)(4)(4) = 1024$

❸ If $x = 3$ and $y = 4$, evaluate x^y. $x^y = 3^4 = (3)(3)(3)(3) = 81$

Evaluate a _factorial_: The form of a factorial is $n!$ To evaluate it, multiply n by $(n - 1)$, then by $(n - 2)$ and so on until you get to "1". If there are multiple factorials, evaluate each one separately first and then complete any other operations that are required to obtain a final result.

Examples

❶ $6! = (6)(5)(4)(3)(2)(1) = 720$

❷ $4! \, 5! = (4)(3)(2)(1) \bullet (5)(4)(3)(2)(1) = 2880$

❸ $\dfrac{6!}{3!} = \dfrac{6 \bullet 5 \bullet 4 \bullet 3 \bullet 2 \bullet 1}{3 \bullet 2 \bullet 1} = \dfrac{720}{6} = 120$

10 – FACTORING

DEFINITIONS

<u>Factors</u>: Numbers, terms, or expressions that are multiplied together to form a product. A polynomial inside a ( ) is a factor if the ( ) indicates multiplication, not just grouping.

Examples

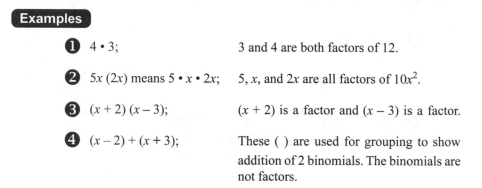

❶ $4 \cdot 3$;   3 and 4 are both factors of 12.

❷ $5x\,(2x)$ means $5 \cdot x \cdot 2x$; 5, x, and $2x$ are all factors of $10x^2$.

❸ $(x + 2)\,(x - 3)$; $(x + 2)$ is a factor and $(x - 3)$ is a factor.

❹ $(x - 2) + (x + 3)$;   These ( ) are used for grouping to show addition of 2 binomials. The binomials are not factors.

<u>Factoring A Problem</u>: Breaking down a problem into its prime factors.

<u>Factor Completely</u>: When an expression is completely broken down into its prime factors. There will only be one correct set of prime factors for any expression.

<u>Prime Factors</u>: These are the factors of a product that are broken down as far as possible while still resulting in the same product. The prime factors of a number or of an algebra problem will always be the same.

Examples

❶ $12 = 2 \cdot 6$; $12 = 2 \cdot 2 \cdot 3$;   2, 2, and 3 are all prime factors of 12.

❷ $3x + 6 = 3\,(x + 2)$;   3 and $(x+2)$ are the prime factors of $3x+6$.

❸ $12xy + 3x = 3x\,(4y + 1)$;   3 and x and $(4y + 1)$ are prime here.

Common Factor (GCF): A factor which is present in each term in an expression. The greatest common factor or GCF is the largest factor that is present in each of the terms to be considered. Each term in the expression must be divisible by the same number(s) and/or variable(s) if a GCF exists. A GCF greater than one does not always exist. If it does, we "factor it out" which means to divide each term in the expression by the GCF. It is kept with the other factors.

Examples

❶ GCF of $4x + 8y$ is 4; $4x$ and $8y$ are both divisible by 4.

Factors are $4(x + 2y)$

❷ GCF of $x^2 + 2x$ is x; x^2 and $2x$ are both divisible by x.

Factors are $x(x + 2)$

❸ GCF of $12xy - 4xyz$ is $4xy$; $12xy$ and $-4xyz$ are both divisible by $4xy$.

Factors are $4xy(3 - z)$

Procedure for factoring binomials and trinomials:

Steps:

1) In any factoring problem, put the terms in standard form first.

Note: In an *equation*, make one side of the equation $= 0$ by moving all the variables and numbers to one side of the equal sign. (Use the usual algebraic methods of adding or subtracting terms from both sides of the equation.)

2) LOOK at each term in the problem to see if there is a GCF. If there is, factor out the GCF and show it at the left side of the remaining expression. The GCF remains as part of the problem. Go to step 3. If there is no GCF, go directly to step 3.

3) Now look INSIDE the parenthesis, or just at the problem itself if there was no GCF, to see if what is left can be factored. Is it a binomial or trinomial? (see next page to factor a binomial, and see page 31 if it is a trinomial)

Examples

❶ $2x^2 + 6x - 8$ GCF is 2

$2(x^2 + 3x - 4)$ Show the 2 at the left, and the quotient after "factoring out" the 2 in parenthesis.

❷ $x^2 + 2x - 15$ No GCF. Go to step 3

❸ $2x^3 - 8x$ GCF is $2x$

$2x(x^2 - 4)$ $2x$ and the quotient are both shown.

FACTORING BINOMIALS

The Difference of two perfect squares, like $x^2 - a^2$, can be factored into $(x - a)(x + a)$. When $(x - a)(x + a)$ are multiplied together the result is $x^2 - a^2$.

Examples

1 $x^2 - 9$
$(x + 3)(x - 3)$

This expression has no GCF. It is the difference of 2 perfect squares.

2 $2x^3 - 8x$
$2x(x^2 - 4)$
$2x(x - 2)(x + 2)$

This expression has a GCF, and the quotient is the difference of 2 perfect squares.

3 $4x^2 - 81$
$(2x + 9)(2x - 9)$

In this example, the first term has a perfect square number as well as a squared variable. The factors are shown.

The Sum of two perfect squares, like $x^2 + a^2$. This is prime and cannot be factored.

Examples

1 $x^2 + 16$

Both terms are perfect squares and they are added. This cannot be factored.

2 $2x^2 + 18$
$2(x^2 + 9)$

There is a GCF of 2. Factor that out and the quotient that remains is prime.

Some binomials are not factorable - some have a GCF and that is all that can be factored, others are prime as they are.

Examples

1 $3x^2 + 6$
$3(x^2 + 2)$

GCF of 3
Cannot be factored further.

2 $5x^2 - 4$

Examples 2, 3, and 4 cannot be factored at all.

3 $x^2 + 1$

4 $x^2 - 2$

<u>Factoring Trinomials into 2 binomial factors</u>: When working with a trinomial in standard form, we use "first", "middle", and "last" when referring to the position of the term. For example, in the trinomial: $5x^2 + 3x - 2$, $5x^2$ would be the first term, $3x$ the middle term, and -2 the last term and is also called the constant. As a reminder, the standard form of a quadratic trinomial is $ax^2 + bx + c$. The numbers represented by a, b, and c, can each be negative or positive real numbers.

The *leading coefficient, "a"*, is the numerical factor of the first term. In the example above, $5x^2$ is the first term which makes 5 the leading coefficient. In Algebra 1 we will be factoring trinomials with a leading coefficient of "1" after a GCF is factored out if possible. When written in standard form, a quadratic trinomial's leading coefficient is usually referred to as "*a*".

> *Note*: If the leading coefficient is one, it is not written in front of the first term. In algebra, the number one is rarely written down when it is used to multiply something. x^2 means $(1)(x^2)$

The coefficient of the 2nd term, the x term, is called "*b*". Again, if it is one, it is not written down. x means $(1)(x)$

The constant, or the third term which is a number only, is called "*c*." If c is 1, then "1" IS written here because it is not a factor but is simply a number.

Follow these steps to successfully factor completely a quadratic trinomial into its binomial factors.

> *Note*: Not all quadratic trinomials can be factored and we have other methods of working with them that will be presented at a later time.

1. Make sure the trinomial is in standard form, $ax^2 + bx + c$, and that a GCF has been factored out if there is one. Remember that the GCF stays in the problem.

2. Identify the values of b and c, including their + or – signs. On scrap paper, write down all the pairs of factors (numbers) that multiply together to give a product equal to "*c*".

3. Find the sum (add them together) of each pair of factors for "*c*" that you found in step 2 and identify the pair that has a sum equal to "*b*" - making sure the sum has the correct sign. Only one pair will work correctly.

4. Underneath the original trinomial, make 2 sets of parentheses, side by side. (Each parenthesis will contain a binomial factor when we are finished.)

5. In both parentheses, put x as the first term. The 2nd terms in the each parenthesis will be the pair of factors (from step 4) of "*c*" whose sum equals "*b*" (from step 5). Include the positive or negative signs.

6. You can check to make sure the factors are correct by multiplying the 2 binomials back together. If a GCF was found, make sure to multiply by it back in using the distributive property as the last step. The original trinomial should be the result.

Here are some examples: There are more on the next page.

Notice that in examples 1 and 2, the sign of the *last term, c, is positive*. In order for *c* to be positive, its factors must both be negative or both positive. Therefore, the signs in both *parentheses will be alike* and match the sign of the 2nd term, "*b*".

Examples

❶ $x^2 + 10x + 24$ $b = +10$ and $c = +24$.
    $(x + 6)(x + 4)$ Factors of 24 are (+2)(+12), (+3)(+8), (+6)(+4),
                          (+24)(+1), (–2)(–12), (–3)(–8), and (–6)(–4), (–24)(–1)
                          (+6) and (+4) are the only pair that adds up to +10 and
                          they are the factors needed.

❷ $n^2 – 14n + 24$ $b = –14$ and $c = +24$.
    $(n – 12)(n – 2)$ Factors of 24 are (+2)(+12), (+3)(+8), (+6)(+4),(+24)(+1),
                          (–24)(–1)(–2)(–12), (–3)(–8), and (–6)(–4). Same as above!!
                          (–12) and (–2) are the only two that add up to –14.

In examples 3 and 4, the sign of the *last term, c, is negative*. Its factors must be one positive and one negative. The signs in the parentheses will be *one positive and one negative*. Only one pair of factors of *c* will have the correct sum with the correct sign!

❸ $y^2 + 6y – 27$ $b = +6$, $c = –27$
    $(y + 9)(y – 3)$ Factors of –27 are (+3)(–9), (–3)(+9), (+27)(–1), (–27)(+1)
                          The only pair that has a sum of + 6 is (+9) and (–3)

❹ $x^2 – 3x – 18$ $b = –3$ and $c = –18$
    $(x + 3)(x – 6)$ Factors of –18 are (+18)(–1), (–18)(+1), (+6)(–3),
                          (–6)(+3), (+9)(–2) and (–9)(+2)
                          (–6) and (+3) have a sum of –3 and are the correct factors.

Note: The order of the parentheses doesn't matter. In example 4 we could have had $(x – 6)(x + 3)$ and that would be perfectly OK. We must use –6 and +3 in the parentheses, not –3 and + 6. The important part is to make sure that the numbers have the correct sign.

Leading Coefficient of –1: We don't want to factor with a negative value for the first term. Use (–1) as a GCF and when it is factored out, it will change all the signs to make the problem manageable. See example 9, next page.

Factoring Examples Several examples follow for review. It is not possible to show all the possibilities in factoring.

Examples

❶ $4x^3 + 6x$ factors as $2x(2x^2 + 3)$ GCF and it cannot be factored further.

❷ $x^2 + x$ factors as $x(x + 1)$ GCF and it cannot be factored further.

❸ $2x^2 - 32 = 2(x^2 - 16) = 2(x - 4)(x + 4)$ GCF, then difference of perfect squares.

❹ $3x^2 + 27 = 3(x^2 + 9)$ GCF, then the SUM of 2 perfect squares has no factors.

❺ $x^2 + 5x + 6 = (x + 3)(x + 2)$ All signs are positive.

❻ $x^2 - 9x + 20 = (x - 5)(x - 4)$ Last sign (c) is +, Middle sign (b) is –.

❼ $x^2 - 3x - 28 = (x - 7)(x + 4)$ Last sign (c) is –, One ( ) has +, the other has –.

❽ $x^2 + 9x - 36 = (x + 12)(x - 3)$ Last sign (c) is –, One ( ) has +, the other has –.

❾ $-x^2 + x + 2 = (-1)(x^2 - x - 2) = (-1)(x - 2)(x + 1)$ Leading coefficient is negative one. Fix that by using (-1) as the GCF. Then proceed with the factoring as usual.

❿ $2x^2 - 10x + 12 = 2(x^2 - 5x + 6) = 2(x - 3)(x - 2)$ GCF is 2. The trinomial that is left after factoring the GCF is factorable.

Check $2(x^2 - 5x + 6) = 2x^2 - 10x + 12$ To check, multiply the two binomials and then multiply by the GCF, 2.

⓫ $x^2 + 3x + 4$ Cannot be factored.

Remember to check your factors by multiplying them back together. Multiply the binomials first, simplify, then use the distributive property if you have a GCF. See example number 9 and 10 above, to demonstrate the check.

Example

Steps: **1)** $(-1)(x - 2)(x + 1)$ Factors
Multiply the binomials. You can use FOIL.

2) $(-1)(x^2 + x - 2x - 2)$ Use the Distributive Property with (-1)

3) $(-1)(x^2 - x - 2)$ The product is the same as the original example number 9 and 10 above.

4) $-x^2 + x + 2$

11 – ALGEBRAIC FRACTIONS

An algebraic fraction contains one or more variables. The same rules apply to performing operations with algebraic fractions as apply to fractions containing only numerals.

"Cancel" is a commonly used student friendly math word that is not always accepted by teachers. It means to divide each term of the numerator and each term of the denominator of a fraction by any factor(s) common to both. When canceling results in "1" make sure and write it down. If the numerator or denominator contain binomials and/or trinomials it is a good idea to use parentheses to group the terms. Only factors can be cancelled - not individual terms.

Simplifying or Reducing: Factor the numerator and the denominator completely. Then divide each term of the numerator and the denominator by all the factors that are common to each of the terms.

Examples

❶ $\dfrac{(3x+6)}{3} = \dfrac{\cancel{3}(x+2)}{\cancel{3}(1)} = \dfrac{x+2}{1} = x+2$ 3 is the common factor.

❷ $\dfrac{(x^2+2x)}{6x} = \dfrac{\cancel{x}(x+2)}{\cancel{x}(6)} = \dfrac{x+2}{6}$ x is the common factor

❸ $\dfrac{3x}{6x^2} = \dfrac{\cancel{3x}(1)}{\cancel{3x}(2x)} = \dfrac{1}{2x}$ $3x$ is the common factor.

❹ $\dfrac{3xy^2}{3xy^2} = \dfrac{\cancel{3xy^2}(1)}{\cancel{3xy^2}(1)} = \dfrac{1}{1} = 1$ Everything "cancels". The answer is 1.

❺ $\dfrac{(n^2+2n)}{(n^2+3n+2)} = \dfrac{n\cancel{(n+2)}}{\cancel{(n+2)}(n+1)} = \dfrac{n}{(n+1)}$ $(n+2)$ is the common factor.

In example number 5, one factor in the binomial numerator cancels with one factor of the trinomial denominator. Parenthesis were added to group the numerator terms and the denominator terms. This avoids an error of trying to cancel the "2" in the numerator with the 2 in the denominator.

Examples of Simplifying or Reducing (continued)

$$\textbf{6} \quad \frac{(x^2-1)}{(3x^2-3x)} = \frac{(x+1)\cancel{(x-1)}}{3x\cancel{(x-1)}} = \frac{(x+1)}{3x}$$

In example number 6, the binomial numerator is the difference of 2 perfect squares. One of its factors cancels with the binomial factor in the denominator after the GCF of $3x$ is factored out of the denominator. Again - parentheses helps to avoid errors in cancelling. The final answer can be left with the ( ) or they can be removed.

$$\textbf{7} \quad \frac{x^2-6x-72}{x^2-8x-84} = \frac{(x-12)\cancel{(x+6)}}{(x-14)\cancel{(x+6)}} = \frac{(x-12)}{(x-14)}$$

In example number 7, the original fraction has 2 trinomials to be factored before any simplifying can be completed.

Note: The final answer contains a binomial in the numerator and in the denominator. Nothing further can be cancelled.

$$\textbf{8} \quad \frac{x-2y}{4y-2x} = \frac{(x-2y)}{(-2)(-2y+x)} = \frac{\cancel{(x-2y)}}{(-2)\cancel{(x-2y)}} = \frac{1}{-2} \ \text{or} \ -\frac{1}{2} \ \text{or} \ \frac{-1}{2}$$

In example number 8, the numerator and denominator seem alike but the signs of the numerator are opposite the signs in the denominator. Re-arrange the denominator by factoring out a -2. That changes the signs on the denominator and it can be rewritten to match the numerator. Be careful to keep the -2 in the problem! Then cancel and simplify to a final answer of $-1/2$. Reminder: The 3 choices shown for the answer are equivalent fractions.

$$\textbf{9} \quad \frac{x^2+x-2}{x^2-x-2} = \frac{(x+2)(x-1)}{(x-2)(x+1)}$$

In example number 9, the fraction cannot be simplified. None of the factors can be cancelled. (Changing the signs by factoring out -1 is not helpful because the x^2 terms are both the same).

Fractions with signs: These negative fractions are all equivalent and can be used in whichever form is needed in the problem.

Examples   **1** $\dfrac{-5x}{10y} = -\dfrac{5x}{10y} = \dfrac{5x}{-10y}$

2 $-\dfrac{2(x-1)}{x+4} = \dfrac{-2(x-1)}{x+4} = \dfrac{2(x-1)}{-(x+4)}$

Equivalent Fractions: 2 or more fractions that have the same value when they are reduced to lowest terms.

Examples   **1** $\dfrac{2}{3} = \dfrac{4}{6}$ $\dfrac{4}{6}$ is $\dfrac{2}{3}$ when it is reduced.

2 $\dfrac{5x}{10y} = \dfrac{10x}{20y} = \dfrac{x}{2y}$

Equivalent fractions can be made by multiplying the numerator and denominator of a fraction by the same factor. (This process is needed to add and subtract fractions.)

Examples   **1** $\dfrac{3}{5} = \dfrac{(2)(3)}{(2)(5)} = \dfrac{6}{10}$ The numerator and denominator are multiplied by 2.

2 $\dfrac{2x}{7xy} = \dfrac{(2y)(2x)}{(2y)(7xy)} = \dfrac{4xy}{14xy^2}$ The multiplying factor here is $2y$.

Addition or Subtraction: A common denominator is needed in all fractions that are added or subtracted. The initials LCD are used to show Least Common Denominator. (Least means smallest in this description.)

Step: **1)** Make equivalent fractions with a common denominator.

2) Put a parenthesis around the numerator of a fraction that is subtracted.

3) Make one big fraction with the numerators all written together and one denominator for the whole fraction.

4) Use the distributive property if needed.

5) Collect like terms.

6) Simplify the fraction.

Examples

① $15y$ is the LCD

$$\frac{2x}{3y} + \frac{4x}{5y} = \frac{(5)(2x)}{(5)(3y)} + \frac{(3)(4x)}{(3)(5y)} = \frac{10x+12x}{15y} = \frac{22x}{15y}$$

② $15x$ is the common denominator in this subtraction example.

$$\frac{2x}{3x} - \frac{4x}{5} = \frac{5(2x)}{5(3x)} - \frac{3x(4x)}{3x(5)} = \frac{5(2x)-3x(4x)}{15x} = \frac{10x-12x^2}{15x}$$

③ The LCD here is $(x-5)$. Add the numerators, keep the LCD.

$$\frac{2x}{x-5} + \frac{x-6}{x-5} = \frac{2x+x-6}{x-5} = \frac{3x-6}{x-5}$$

④ The LCD is $(x+2)$. Put a ( ) around the numerator of the 2nd fraction. Combine the numerators into one numerator showing the subtraction of the $(3x+4)$. Keep the LCD as it is. Distribute the $-$ sign, collect like terms.

$$\frac{7x}{x+2} - \frac{(3x+4)}{x+2} = \frac{7x-(3x+4)}{x+2} = \frac{7x-3x-4}{x+2} = \frac{4x-4}{x+2}$$

⑤ The LCD is $(3x+1)$ and the problem is addition. Combine the 2 numerators into one fraction, keep the LCD. Notice that the numerator after like terms are collected can be factored and the fraction can be reduced by cancelling.

$$\frac{(2x+4)}{3x+1} + \frac{(4x-2)}{3x+1} = \frac{(2x+4)+(4x-2)}{3x+1} = \frac{6x+2}{3x+1} = \frac{2\cancel{(3x+1)}^{1}}{\cancel{(3x+1)}_{1}} = \frac{2}{1} = 2$$

⑥ After combining the numerators, notice that the difference of 2 perfect squares is the result. Factor it and then cancel the $(x+3)$ in the numerator and in the denominator.

$$\frac{x^2}{x+3} - \frac{9}{x+3} = \frac{x^2-9}{x+3} = \frac{(x-3)\cancel{(x+3)}}{\cancel{(x+3)}} = \frac{x-3}{1} = x-3$$

⑦ The LCD of $(x+2)$ is carried throughout the problem. When the numerators are combined and like terms are collected, a quadratic trinomial which can be factored is the result. Factor and then cancel the $(x+2)$ in the numerator and the denominator to reduce the fraction.

$$\frac{(2x^2-6)}{x+2} - \frac{(x^2+x)}{x+2} = \frac{(2x^2-6)-(x^2+x)}{x+2} = \frac{2x^2-6-x^2-x}{x+2} = \frac{x^2-x-6}{x+2} = \frac{(x-3)\cancel{(x+2)}}{\cancel{(x+2)}} = x-3$$

Multiplication and Division of Fractions: A common denominator is not needed.

Multiplication: 1. Cancel where possible. If the numerator and/or denominator is a binomial or trinomial, it may be necessary to factor each before canceling. Canceling can be done between 2 fractions as long as there is a factor in the denominator and factor in the numerator. (You can't cancel using 2 denominators or 2 numerators together.)

2. Multiply the numerators together.

3. Multiply the denominators together.

4. Simplify if needed. (Good cancelling usually results in an already simplified answer.)

Examples

❶ $\dfrac{14x}{5} \cdot \dfrac{3y}{7} = \dfrac{\cancel{7}(2x)}{5} \cdot \dfrac{3y}{\cancel{7}(1)} = \dfrac{(2x)(3y)}{5(1)} = \dfrac{6xy}{5}$

❷ $\dfrac{3x}{5y} \cdot \dfrac{2x}{5y} = \dfrac{(3x)(2x)}{(5y)(5y)} = \dfrac{6x^2}{25y^2}$

❸ $\dfrac{x}{(x+2)} \cdot \dfrac{(x-5)}{(x-3)} = \dfrac{x(x-5)}{(x+2)(x-3)} = \dfrac{x^2-5x}{x^2-x-6}$

❹ $\dfrac{x^2-4x+3}{x^2-2x-8} \cdot \dfrac{x^2+5x+6}{(x-3)} = \dfrac{(x-1)\cancel{(x-3)}}{\cancel{(x+2)}(x-4)} \cdot \dfrac{(x+3)\cancel{(x+2)}}{\cancel{(x-3)}} = \dfrac{(x-1)(x+3)}{(x-4)} = \dfrac{x^2+2x-3}{x-4}$

Division: 1. Change the fraction after the division sign into its reciprocal (See page 2) by inverting it (flipping it upside down).

2. Change the division sign to a multiplication sign.

3. Multiply the fractions as shown above.

❶ $\dfrac{2x}{5} \div \dfrac{4}{3} = \dfrac{2x}{5} \cdot \dfrac{3}{4} = \dfrac{(2x)(3)}{(5)(4)} = \dfrac{\cancel{2}(x)(3)}{5(\cancel{2})(2)} = \dfrac{3x}{10}$

❷ $\dfrac{(x+2)}{2x} \div \dfrac{(x-1)}{4x} = \dfrac{(x+2)}{\underset{1}{\cancel{2x}}} \cdot \dfrac{\overset{2}{\cancel{4x}}}{(x-1)} = \dfrac{2(x+2)}{(x-1)} = \dfrac{2x+4}{(x-1)}$

Note: Make sure to change to multiplying by the reciprocal <u>before</u> canceling anything when doing division.

Problems containing whole numbers, mixed numbers, and fractions: Any whole number can be made into a fraction by simply putting a "1" as a denominator and then treating it like a fraction. Mixed numbers are usually changed to improper fractions before working with them. In algebra, we rarely see mixed numbers.

Decimals and Fractions: If a problem is given in fraction form, then it should be kept in fraction form in most cases. A fraction is considered an "exact" answer while a decimal may not be. Problems that are given in decimal form to begin with are usually kept in decimal form. If the problem has both fractions and decimals in it, then you teacher may have a preference about which form you use -- so ask!

Equations containing fractions -- see page 54
Fractions as ratios and proportions -- see page 44

FRACTION SUMMARY — ALGEBRAIC AND NUMERIC FRACTIONS

Operations	Common denominator needed? LCD	Cancel?	Procedure
$+ / -$	Yes	Only if one fraction can be reduced before finding an LCD.	Add or subtract the numerators. Keep the new denominator.
$\times$	No	Yes–Factor First. Cancel within one fraction or cancel a factor in the numerator of the fraction with a factor in the denominator of another.	Multiply numerators. Multiply denominators. Simplify if needed.
$\div$	No	Yes–First invert (flip) the fraction after the $\div$ sign. Then cancel the same way as in multiplication of the fraction.	"Flip" the fraction after the $\div$ sign. Cancel if possible. Use multiplication procedure.
Equations			
Proportions 2 = fractions (ratios) $\dfrac{3x}{12} = \dfrac{6}{24}$	No	ONLY within each fraction. *Do not* cancel $\dfrac{\overset{1}{\cancel{3x}}}{\underset{4}{\cancel{12}}} = \dfrac{3}{4}$	Cross multiply. $\dfrac{x}{4} \diagdown \dfrac{3}{4}$ "across" the equal sign. $4x = 12$ $x = 3$ Solve for the variable.
Equations with sums or differences of fractional terms $x + \dfrac{x+1}{10} = \dfrac{13}{20}$	Yes LCD is 20	Cancel while "clearing" the fractions. To "clear" multiply each term in the equation by the LCD. Cancel where possible.	$20(x) + \overset{2}{\cancel{20}}\left(\dfrac{x+1}{\underset{1}{\cancel{10}}}\right) = \overset{1}{\cancel{20}}\left(\dfrac{13}{\underset{1}{\cancel{20}}}\right)$ $20x + 2(x+1) = 13$ $20x + 2x + 2 = 13$ $22x + 2 = 13$ $22x = 11$ $x = \dfrac{1}{2}$

Examples of fractions with binomial or trinomial denominators are shown on the pages for adding/subtracting and multiplying/dividing. Remember that a fraction means division and that division by zero is undefined. Any work we do with fractions must have a denominator that is not equal to zero.

12 – SIMPLIFYING RADICALS

Index

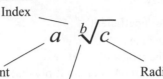

$$a \quad \sqrt[b]{c}$$

Coefficient   Radicand

Radical Sign

A **radical sign** $\left(\sqrt{}\right)$ in a problem indicates that a root of the number under the "radical" is to be found. If no index is given, then the *positive square root* of the radicand is the answer.

Example $\sqrt{81} = 9$

We know that the square root of 81 could also be (– 9), but the radical sign indicates use of the positive root only. If both square roots of a number are to be used, a ± sign is in front of the radical.

Example $\pm\sqrt{81} = \pm 9$ or $+9$ and -9

When solving a quadratic equation (see page 82) we have to indicate that the positive and negative roots are to be used.

Example *If* $x^2 = 25$, *then* $x = \pm\sqrt{25}$, *and* $x = \pm 5$

Note: Although all positive numbers have both a positive and a negative square root, we will refer to positive square roots in this section unless directed otherwise. When finding a square root, *if the radicand is a negative number, there is no real square root.*

Root: A root is a factor of the radicand which is multiplied by itself a given number of times to produce the number under the $\sqrt{}$. A square root means to multiply the answer by itself to get the radicand. A cube root, used as a factor three times has a product equal to the radicand.

Example the square root of 81 is 9 because 9 • 9 = 81

Radicand: The number (and/or variables) under the radical sign.

Index: The small number over the left edge of the radical sign which indicates the type of root to be found. If there is no number there, it is assumed to be "2".

Examples

  ❶ $\sqrt[3]{8} = 2$ 3 is the index. Find the cube root of 8 which is 2.

  ❷ $\sqrt{16}$ There is no index so an index of 2 is understood. Find the square root of 16 which is 4.

Integrated Algebra Made Easy

   41

Coefficient: Refers to the number outside the radical sign. Since radical signs are often treated as variables, "coefficient" seems appropriate. Multiply the root by the coefficient for a final answer.

Example $7\sqrt[3]{64}$ 7 is the "coefficient"
                        ^{3} is the index
                        64 is the radicand
In words: Find the cube root of 64, then multiply that number by 7. The cube root of 64 is 4. The final answer is $7(4) = 28$.

"Like" Radicals: Have the same index and the same radicand.

Example $\sqrt{5}$ *and* $2\sqrt{5}$ are "like." $\sqrt{2}$ and $\sqrt{3}$ are not "like radicals."

Rational and Irrational Roots:
1. The square root of a positive number that is a *perfect square* is a *rational* number. No $\sqrt{}$ will be visible in the answer.

Example $\sqrt{36} = 6$

2. The square root of any positive number that is *not a perfect square* is an *irrational* number. Irrational square roots are left "in simplified radical form" unless directed otherwise and a radical sign will be part of the answer.

Examples ❶ $\sqrt{8} = 2\sqrt{2}$ in simplest radical form. (This is an EXACT answer.)

❷ $\sqrt{8} \approx 2.8284271$ Write the full display of the calculator on your paper unless directed to round.

EXACT answers require that a root that is irrational be left in radical form. If an estimated answer is needed, use the $\sqrt{}$ button on your calculator and then round the answer to the place indicated in the problem.

Simplifying Radicals: First test the number under the radical to see if it is a perfect square. Use the $\sqrt{}$ button on your calculator - if the answer is a whole number, you are done! If not - follow this procedure to simplify:
1. Break the number under the radical sign (the radicand) into its prime factors.
2. Remove from the radical any *pair* of numbers that are factors - placing one of each pair on the outside of the radical sign. The "other one" of the pair of numbers disappears because by removing the pair from the radical, you have actually found the square root of that pair of factors.
3. Continue this process until as many pairs of factors - or "perfect squares" as possible are removed from under the $\sqrt{}$.

(continue on the next page)

Integrated Algebra Made Easy

4. The outside factors are multiplied together to make the "coefficient" of the radical.
5. Multiply back together any factors remaining under the radical sign.

Example $\sqrt{540} = \sqrt{\underbrace{3 \cdot 3} \cdot 3 \cdot \underbrace{2 \cdot 2} \cdot 5} = 3 \cdot 2\sqrt{3 \cdot 5} = 6\sqrt{15}$

Simplifying with variables: Use the same process shown above, but include the variables in the factoring process.

Example $\sqrt{12x^3y^2} = \sqrt{\underbrace{2 \cdot 2} \cdot 3 \cdot \underbrace{x \cdot x} \cdot x \cdot \underbrace{y \cdot y}} = 2 \cdot x \cdot y\sqrt{3 \cdot x} = 2xy\sqrt{3x}$

This example has a pair of 2's, a pair of x's and a pair of y's. Find the square root of each pair and move it to the outside of the radical sign. Multiply the outside factors together. Then multiply the inside factors together.

Adding or Subtracting Radicals:
1. Only "like" radicals can be added.
   Sometimes it is possible to make them "alike" by simplifying the original problem. Both examples below are shown being simplified first.
2. Add (or subtract) the coefficients.
3. Keep the radicand unchanged.

Examples ❶ $\sqrt{8} + \sqrt{32} = 2\sqrt{2} + 4\sqrt{2} = 6\sqrt{2}$

❷ $3\sqrt{50x} - 85\sqrt{2x} = 15\sqrt{2x} - 85\sqrt{2x} = -70\sqrt{2x}$

4. Some radicals cannot be added or subtracted. Simplify first, as usual. When it becomes evident that they cannot be simplified to match, leave the answer as shown here. The answers will have several simplified parts.

Examples ❶ $\sqrt{15} + \sqrt{20} = \sqrt{15} + 2\sqrt{5}$ This one has 2 radicals for the answer.

❷ $\sqrt{25} - \sqrt{60} = 5 - 2\sqrt{15}$ This one has a whole number and a radical for the answer.

Multiplying or Dividing Radicals:
1. The radicands do not have to match.
2. Multiply or divide the coefficient of one by the coefficient of the other.
3. Multiply or divide the radicands. Notice that when working on division problems both radicands are put into one radical sign (see example #3 below).
4. Simplify if possible.

Examples ❶ $\sqrt{5} \cdot 6\sqrt{10} = 6\sqrt{5 \cdot 10} = 6\sqrt{50} = 30\sqrt{2}$ ❸ $\frac{4\sqrt{12}}{2\sqrt{6}} = \frac{4}{2}\sqrt{\frac{12}{6}} = 2\sqrt{2}$

❷ $x\sqrt{3y} \cdot 4x\sqrt{5z} = 4x^2\sqrt{15yz}$ ❹ $\frac{24\sqrt{6}}{3} = \frac{24}{3}\sqrt{6} = 8\sqrt{6}$

13 – RATIO AND PROPORTION

Ratio: A comparison of two numbers. Ratios can compare numbers themselves or they can compare two numbers with different units.
Ratios are usually simplified - written in lowest terms.
There are several ways to write ratios:
Numbers: 4:6, or $\frac{4}{6}$. "4 is to 6" is often used to describe a ratio in a proportion.
Numbers with different units: 20 miles: one gallon, or 20 miles/gallon, or 20 miles per gallon. 60 students: 2 teachers becomes 30 students: 1 teacher, or 30 students/teacher, or 30 students per teacher.

Proportion: 2 equal ratios. These are used in equations. Solve equations that are proportions by cross multiplying. Proportions are used in scale drawings, maps, shadow problems, and many other kinds of problems.

Examples

❶ $\dfrac{x}{3} = \dfrac{5x+2}{18}$; $18(x) = 3(5x+2)$; $18x = 15x + 6$; $3x = 6$; $x = 2$

❷ $\dfrac{x^2 + 2}{20} = \dfrac{2x+1}{10}$

$10(x^2 + 2) = 20(2x + 1)$

$10x^2 + 20 = 40x + 20$

$10x^2 - 40x = 0$

$10x(x - 4) = 0$

$10x = 0 \quad x - 4 = 0$

$x = 0 \qquad x = 4$

his proportion becomes a quadratic equation when it is cross - multiplied. It must be factored and then, using the multiplication property of zero, solved. See Chapter 19 —
Quadratic Equations, page 84.

Percent: Use $\dfrac{is}{of} = \dfrac{\%}{100}$ or $\dfrac{part}{whole} = \dfrac{\%}{100}$ when setting up the problem.

Examples

❶ 15 is 10 % of what number?

$15/x = 10/100$

$10x = 1500$

$x = 150$

> **_Note_:** / is used as a fraction line.
> 10/100 means ten-hundreths.

❷ What percent of a 12 piece pizza is 9 pieces?

$9/12 = x/100 \%$

$900 = 12x$

$x = 75\%$

PERCENTS – INCREASE, DECREASE, AND DISCOUNT

Discount: Stores often give a % off as a discount %. Find the sale price by using this formula: *Original selling price – Discount amount = Discount sale price*

Note: *The dollar sign is usually omitted in the formula but must be put back in the answer when the answer is in dollars.*

Examples

1 Jerome bought a pair of sneakers at a 25% discount sale. He paid $60.00. What was the original price of the sneakers?

Original Price $= x$
Discount amount $= 0.25x$ (Change the % to a decimal.)
Sale Price $= \$60$
Substitute in the formula: $x - 0.25x = 60$
$$0.75x = 60$$
$$x = 80 \quad \text{The original price was } \$80.$$

2 Find the cost of a $200 bike that is on sale at 15% off.

Original Price $= \$200$, Discount Amount $= (0.15)(\$200)$, Discount sale price $= x$
$200 - (0.15)(200) = x$ Substitute
$\quad\quad 200 - 30 = x$ Solve for x
$\quad\quad\quad\quad x = 170$ The discount sale price of the bike is $170.

3 What is the discount on a skateboard that was originally sold for $150 and is now on sale for $120?

Original Price $= \$150$, Discount Amount $= (x)(\$150)$, Discount sale price $= \$120$
$150 - 150x = 120$ Substitute
$\quad 150x = 120 - 150$
$\quad\quad x = \dfrac{30}{150}$ Solve
$\quad x = 0.2$ (Change to a % by multiplying by 100%)
$\quad\quad\quad$ The discount is 20% off.

Decrease: Another way to say this is that the sale price is a *decrease of 20%* from the original price. This is called a **percent of decrease.**

(Examples continued from previous page.)

4 My pool was 72°F on Monday. On Friday it was 80°F. What is
the **percent of increase**, to the nearest tenth, of the temperature in the pool?
This is not a money problem, but it can be handled in a similar way. Since
the temperature is increasing, use + the change for the 2nd term.

Original + change = final result
Original temperature = 72, % change = x, Final temperature = 80

$$72 + (x)(72) = 80$$
$$72x = 8$$
$$x = 0.1111... \quad \text{Change to } \% = 11.111...\%$$
The pool temperature increased by about 11.1%

Similarity: 2 polygons are similar if they have corresponding sides that are
proportional and corresponding angles that are congruent. The order in which
the letters identifying the vertices in one polygon are written to match the order of
the corresponding vertices of the 2nd polygon. To find the lengths of sides of
similar polygons, we use proportions.

Solving Geometry problems using proportions:

1. Draw diagrams of both similar figures and label the known sides.

2. Make two equal ratios using information from one polygon for the
   numerators of both fractions and corresponding information from the
   other polygon as the denominators of both fractions.

3. Cross multiply and solve.

ΔABC is similar to ΔRST. If $\overline{BC} = 12$ and $\overline{AB} = 6$,
find the lengths of $\overline{RS}$ and $\overline{ST}$ if $\overline{RS}$ is 2 units less than $\overline{ST}$.

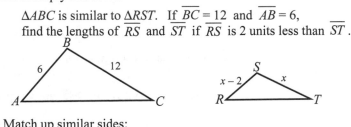

Steps: 1) Match up similar sides:

2) Set up the proportion: $\dfrac{12}{6} = \dfrac{x}{x-2}$

3) Solve for x:
$$12(x - 2) = 6x$$
$$12x - 24 = 6x$$
$$\underline{-12x \qquad -12x}$$

4) Simplify
$$\frac{-24}{-6} = \frac{-6x}{-6}$$
$$4 = x$$

5) Answer: Side $\overline{ST} = 4$, and side $\overline{RS} = 2$

Integrated Algebra Made Easy

Word Problems with Ratios: If a word problem indicates that a ratio is involved, use the ratio numbers to make the "let" statement. In the "let" statement multiply each ratio number by x. Then make the equation and solve as usual.

Examples

❶ Three angles whose sum is 180° have a ratio of 1:2:3.
Find the measures of the angles.

Steps: 1) Assign the unknown: Let x = one angle, $2x$ = next angle, $3x$ = third angle

      2) Set up the equation: $x + 2x + 3x = 180°$

      3) Solve for x: $6x = 180°$
                                   $x = 30°$
      4) Record answer: Therefore the angles are 30°, 60°, and 90°.

❷ Two numbers have a ratio of 3:5. Their sum is 160. Find the numbers.

Steps: 1) Assign the unknown: Let $3x$ = one number, $5x$ = the other number

      2) Set up the equation: $3x + 5x = 160$

      3) Solve for x: $8x = 160$
                                   $x = 20°$
      4) Record answer: The numbers are 3(20) and 5(20) = 60 and 100.

Scale Drawings: A scale drawing is a reduction or an enlargement of a real object. Architectural drawings, models, and maps are some examples of scale drawings.

Scale: The RATIO in the Drawing.

Use the ratio that is the scale of the drawing in a proportion to find information about the drawing or the real object.

General Proportion: $Scale\,(a\ fraction) = \dfrac{drawing}{actual\ or\ real\ object}$

Example

The scale of a map is 1 cm to 5 km. The distance on the map from Santa Fe to Johnstown is 4.5 cm. What is the actual distance between these two cities?

The SCALE is 1 cm : 5 km., the distance in the map or "drawing" is 4.5cm, and we don't know the real object distance.

Steps:

1) Use a variable for the real object distance and make a proportion: $\dfrac{1cm}{5km} = \dfrac{4.5cm}{x}$

2) cross multiply: $1cm \bullet x = (4.5cm)(5km)$

3) divide by 1cm to simplify: (In this step, the "*cm*" cancels.) $\dfrac{1cm \bullet x}{1cm} = \dfrac{(4.5\cancel{cm})(5km)}{1\cancel{cm}}$

4) Answer: $x = 22.5km$

Note: Include the units of measure. Some will cancel while the work is being performed, giving the correct units for the final answer. The actual distance from Santa Fe to Johnstown is 22.5 km.

Direct Variation: Use this formula for direct variation: $y = kx$.
It means that the ratio of y/x is a number which remains constant (unchanged).
1. Find k by dividing y by x.
2. Substitute k in the equation along with whichever variable is given.
3. Find the other variable.

Example   The number of cans of soup packed in 3 cases of soup is 72.
How many cans will be needed to pack 22 cases?

Steps:1)   Put in known values: $72 = k\,(3)$

   2)   solve for k: (24 cans in one case) $\dfrac{72}{3} = k \ \ or \ \ 24 = k$

   3)   Substitute "k" back into equation $y = 24(x)$

   4)   Substitute 22 for x, the known variable: $y = 24\,(22)$

   5)   Solve for y: $y = 528$

   6)   Answer: 528 cans of soup are needed.

Note: Direct variation problems can be solved using proportions. Put the
formula in the form: $\dfrac{y}{x} = \dfrac{k}{1}$ and cross multiply.

Solving Proportions with Inequalities: Whether the proportion involves percent,
money, degrees, etc. and involves a $>$ or $<$ symbol (with or without $=$), solve just
as you would one with $=$ in it except be careful to reverse the inequality symbols
if you have to multiply or divide with a negative number.

Examples

❶ $\dfrac{x}{10} < \dfrac{15}{20}$

    $20x < 150$

    $x < 7.5$

❷ $\dfrac{(x+2)}{-4} \le \dfrac{(x-7)}{2}$

    $2(x+2) \ge -4(x-7)$ Symbol reverses due

    $2x+4 \ge -4x + 28$ to multiplication by (-4)

    $6x \ge 24$

    $x \ge 4$

Many kinds of problems can involve inequalities. They are solved the same way as an
equation would be taking care to reverse the symbols whenever multiplication or
division by a negative number occurs. Make sure to read the problem carefully and
see what kind of answer is required - fraction, decimal, integer, etc.
(See also Chapter 17)

---

14 – DIMENSIONAL ANALYSIS OR CONVERSION FACTORS

Numbers with units can be changed to different units using conversion factors. A conversion factor is a ratio (fraction) that is equal to one. We can multiply the original problem by various combinations of fractions that are equal to one until we get to the units we need.

Example   3 feet/1 yard is a ratio that is equivalent to "1".

Steps:

1) Write the original problem as a fraction with the units included. If it is a whole number, put "1" with the units in the denominator.

2) Plan the conversion factors needed by determining what units the end result contains.

3) Set up the conversion factors (fractions) so that the units will cancel until you get to the units you need. Units in the numerator of one fraction should cancel with units in the denominator of another. Sometimes many factors are needed to get to the end of the problem. Cancel the units until only the units needed for the answer remain in their correct position- numerator or denominator.

4) Multiply all the numerators, then all the denominators, then divide the fraction.

Example   How many inches per minute can a rabbit travel if he can go 2 miles per hour?

$$\frac{2\,\text{miles}}{1\,\text{hour}} \cdot \frac{5,280\,\text{ft}}{1\,\text{mile}} \cdot \frac{12\,\text{inches}}{1\,\text{foot}} \cdot \frac{1\,\text{hour}}{60\,\text{minutes}} =$$

$$\frac{2 \cdot 5,280 \cdot 12 \cdot 1\ \text{inches}}{1 \cdot 1 \cdot 1 \cdot 60\ \text{minutes}} = \frac{2112\ \text{inches}}{1\ \text{minute}} =$$

Answer is: 2112 inches per minute

CONVERSIONS & ABBREVIATIONS

<u>English Measurement System</u>   <u>Abbreviations</u>

   12 inches = 1 foot   12 in = 1 ft *or* 12″ = 1′

   3 feet = 1 yard   3 ft = 1 yd *or* 3′ = 1 yd

   1760 yards = 1 mile   1760 yds = 1 mi

   5280 feet = 1 mile   5280 ft = 1 mi or 5280′ = 1 mi

<u>Metric Measurement System</u>

   10 millimeters = 1 centimeter   10 mm = 1cm

   10 centimeters = 1 decimeter   10 cm = 1dc

   10 decimeters = 1 meter   10 dc = 1 m

   10 meters = 1 dekameter   10 m = 1 dk

   10 dekameters = 1 hectometer   10 dk = 1 h

   10 hectometers = 1 kilometer   10 h = 1 km

Commonly used conversions in the metric system include:

10 millimeters = 1 centimeter   10 mm = 1 cm

100 centimeters = 1 meter   100 cm = 1 m

In the metric system there are also conversion factors that relate length measured in meters, volume or capacity measured in liters, and weight measured in grams.

1cc (cubic centimeter) = 1 gram = 1 milliliter

1000cc = 1 kilogram (kg) = 1 liter (1)

Rates:

Conversions are often used in rate problems.

Rate is defined by this formula: $Rate = \dfrac{Distance}{Time}$

When using this formula it is very important to make sure the units match other parts of the problem - either other information that is given, or the units needed for the solution.

(See examples on next page)

What is Sam's rate if he travels on his bike to a town 15 miles away from home and it takes him 2 hours?

$$R = \frac{D}{T}$$

$$R = \frac{15 \text{ miles}}{2 \text{ hours}}$$

$$R = 7.5 \text{ miles per hour}$$

2 Using the same information as in number one, suppose we want to know the rate in feet per hour instead of the miles per hour as the problem is given. It is necessary to change the units from miles to feet.

$$R = \frac{15 \text{ miles}}{2 \text{ hours}} \bullet \frac{5280 \text{ feet}}{\text{mile}}$$

Change miles to feet by multiplying using 5280 ft/1 mile. The mile units cancel leaving the units of feet which are needed for the answer.

$$R = \frac{(15)5280 \text{ feet}}{2 \text{ hours}}$$

$$R = \frac{79,200 \text{ feet}}{2 \text{ hours}}$$

$$R = 39,600 \text{ feet per hour}$$

3 A snail is crawling along the ground at a rate of 10cm per minute. How many meters will he travel if he crawls for 3 hours?

$$R = \frac{10 \text{ cm}}{1 \text{ minute}}$$

cm needs to be changed to meters, and minutes to hours. Two conversions are needed here.

$$R = \frac{10 \text{ cm}}{1 \text{ minute}} \bullet \frac{1 \text{ meter}}{100 \text{ cm}} \bullet \frac{60 \text{ minutes}}{1 \text{ hour}}$$

cm and minutes both cancel leaving the rate in terms of meters/hour.

$$R = \frac{(10)(60) \text{ meters}}{1(100)(1) \text{ hour}}$$

$$R = \frac{600 \text{m}}{100 \text{hr}} = \frac{6 \text{m}}{1 \text{hr}} = 6 \text{m/hr}$$

Answer: The snail travels 6 meters in one hour so if he travels for 3 hours, he will travel 18 meters.
distance = (rate)(time)

Integrated Algebra Made Easy

15 – SIMPLE EQUATIONS

General Procedure for Solving Simple Equations: Do these steps in the order listed. Use the appropriate steps for a particular problem.

Steps:

1) Remove parenthesis by multiplication (distributive property).

2) Collect like terms on each side of the equal sign.

3) Get all the terms containing the variable on one side of the equal sign and all number terms on the other by adding (add the opposite of what is already in the equation).

4) Separate the variable from its coefficient (multiply or divide by the coefficient).

5) Write the answer as $x = 5$ *or* SS = {5} *or* SS = {x|x = 5}. Circle it.

6) CHECK the answer(s) in the ORIGINAL equation(s). Write the original down and show the substitution of your answer in that equation. Do the arithmetic on both sides of the equation. If the last step shows the two sides of the equation are equal to each other, smile! If not - redo the arithmetic in your check. If *that* is OK, then DO THE PROBLEM AGAIN !!

Equations With One Variable: Isolate the variable (letter). Solve for the variable.

Examples

1 **Steps:** Solve for x: $x + 4 = 12$

1) Get all the number terms on one side and the terms containing x on the other.

$$x + 4 = 12$$

2) Use subtraction to do this.

$$\underline{-4 \quad -4}$$

3) Answer:

$$x = 8$$

2 **Steps:** Solve for y: $5y - 5 = 15$

1) Get all the number terms on one side and the terms containing y on the other.

$$5y - 5 = 15$$

2) Use addition to do this.

$$\underline{+5 \quad +5}$$

3) Isolate the term containing the variable.

$$5y = 20$$

4) Solve for y:

$$\frac{5y}{5} = \frac{20}{5}$$

5) Answer:

$$y = 4$$

Note: Use multiplication or division to do this. The variable can only have "1" as a coefficient when you have solved the equation. [A negative variable $(-x)$ is not an acceptable answer. Change the signs when needed by multiplying both sides of the equation by -1. (x) then becomes positive which is acceptable.]

Decimal Equations: If the problem includes DECIMAL numbers for coefficients, multiply the entire equation by whichever power of 10 is necessary to remove the decimals and make all the numbers in the equation into whole numbers. Then solve as usual.

Example Solve for x: $\qquad\qquad\qquad\qquad\qquad 0.5x + 2 = 17$

Steps: **1)** Remove the decimal and make: $\qquad 10(0.5x + 2) = (17)(10)$

2) Whole numbers: $\qquad\qquad\qquad\qquad 5x + 20 = 170$

3) Solve as usual for x: $\qquad\qquad\quad 5x = 150, \quad x = 30$

Fractional Equation: If the equation contains FRACTIONS, it is best to "clear the fractions" by multiplying the entire equation by the least common denominator. This will remove the denominators and make the fractions into whole numbers. Then solve in the usual way.

Example Solve for x: $\qquad\qquad\qquad\qquad\qquad \dfrac{3}{4}x + 3 = \dfrac{15}{4}$

Steps: **1)** Multiply by 4 to remove the fractions: $(4)\left(\dfrac{3}{4}x + 3\right) = \left(\dfrac{15}{4}\right)(4)$

2) Isolate by subtracting: $\qquad\qquad\quad 3x + 12 = 15$

3) Divide by 3: $\qquad\qquad\qquad\qquad\quad 3x = 3$

4) Answer is: $\qquad\qquad\qquad\qquad\quad x = 1$

Literal Equations: Equations that contain mostly LETTERS instead of numbers. Formulas are literal equations. The directions will usually say, " Solve for t in terms of r and s." or sometimes just, " Solve for m." Solving means to isolate the variable designated. The answer will still have the other letters (and sometimes numbers) in it.

Example $A = lw$ Directions will usually say: Solve for "w" in terms of A and l.

Steps: **1)** Divide both sides by l to isolate w: $\qquad \dfrac{A}{l} = \dfrac{lw}{l}$

2) Answer is: $\qquad\qquad\qquad\qquad\qquad \dfrac{A}{l} = w$

Think of the letter you are solving for as the variable in any ordinary equation. Use the same method you would use to solve a regular equation: Isolate the variable. The answer will contain letters, showing the variable isolated and separated from its coefficient. All the other letters will be on the other side of the equal sign.

Example Solve for x in terms of y: $\qquad\qquad 5(x + 2y) = 3$

Steps: **1)** Use the distributive property: $\qquad 5x + 10y = 3$

2) Add $-10y$ to both sides: $\qquad\quad \underline{-10y \quad -10y}$

$\qquad\qquad\qquad\qquad\qquad\qquad\qquad\qquad 5x = 3 - 10y$

3) Answer is: $\qquad\qquad\qquad\qquad\qquad x = \dfrac{3 - 10y}{5}$

Integrated Algebra Made Easy

WORD PROBLEMS/APPLICATIONS

> **Read all word problems carefully. It easy to make a mistake in the "LET" statement.**

1. Make a "let statement."

2. Translate the words of the problem into algebraic phrases.

3. Solve for the variable.

4. Substitute the value of the variable in the "let" statement to find the other answers (when necessary).

5. Check your answer in the WORDS of the problem.

6. Make a conclusion - answer the question the problem asks - use words. Such as: "The number is 5." *or* "The rectangle is 6 by 7."

7. MAKE SURE your answer is correct for the domain (real numbers unless indicated otherwise). Sometimes an answer must be rejected.

Number Problems: Use x for the number you know least about. Make the other parts of the "let" statement related to that one.

Example   One number is 21 less than twice the other number. Their sum is 54. Find both numbers

Steps: **1)** Let x = one number and let $2x - 21$ = the other number.

2) This is the equation: $\qquad\qquad\qquad\qquad\qquad x + (2x - 21) = 54$

3) Then solve as usual: $\qquad\qquad\qquad\qquad\qquad 3x - 21 = 54$

$$\underline{\qquad\qquad +21 \;\; +21}$$
$$3x = 75$$

4) x is: $\qquad\qquad\qquad\qquad\qquad\qquad\qquad\qquad x = 25$

5) Substitute the value of x to find the other number: $2x-21 \Rightarrow 2(25)-21 = 29$

6) Answer is: $\qquad\qquad\qquad\qquad\qquad\qquad\qquad$ 25 and 29

Special Number Problems - sample "let statements"

Consecutive Integer Problems: Let x = the first consecutive integer, $x + 1$ = the second consecutive integer and $x + 2$ = the third.

Positive or Negative Consecutive Integers: Let statement is still $x, x + 1, x + 2$, etc. Whether + or –, this "let" statement will work for all consecutive integer problems.

Odd or Even Consecutive Integers: Let x = first odd integer, $x + 2$ = next consecutive odd integer. Even integers have the SAME let statement. Again, positive or negative odd/even integers have this same "let" statements. Remember zero is considered an even number.

Coin Problems: There are 2 things to consider in these problems. You must determine whether the problem is giving the NUMBER of coins (how many coins there are) or the VALUE of the coins (what the coins are worth in money). USUALLY THE "LET" STATEMENT tells about the NUMBER OF COINS, NOT THEIR VALUE. The *value is often used in making the equation.* When setting up the equation, if the problem uses dollars then use the decimal values (0.05, 0.10, 0.25) for the coins. If the problem is given in cents, use 5 for nickels, 10 for dimes, 25 for quarters. You may not need values: sometimes the problem asks only about how many coins there are and does not include any information about value.

Example Joe has $2.50. He has 7 more dimes than nickels. How many of each does he have?

Steps: 1) Set up the "Let" statement: Let x = number of nickels

    2) Determine known values: $\therefore$ $x + 7$ = number of dimes

    3) Make equation and insert values of monies: $.05x + .10(x + 7) = 2.50$

    4) Multiply out (distributive property): $.05x + .10x + .70 = 2.50$

    5) Isolate x: $.15x + .70 = 2.50$

$$\frac{\quad\quad\quad -.70 \quad -.70 \quad\quad}{}$$

    6) Solve for x: $\dfrac{.15x}{.15} = \dfrac{1.80}{.15}$

    7) Answer: x = 12 nickels

    8) Plug answer back into "Let" statement: $x + 7$ = 19 dimes

    9) Answer: He has 19 dimes and 12 nickels

Ratio Problems: Word problems sometimes involve ratios between numbers or items. Use the ratio information given in the problem to make the "let" statement. Then use the "let" statement to make an equation for the solution.

Example Find the measure of each angle of a triangle whose angles have a ratio of 3:6:9. (Use information you already know about triangles to form the equation: the sum of the angles in a triangle is 180°.)

Steps: 1) Set up the "Let" statement: Let $3x$ = one angle, $6x$ = the 2nd angle, and $9x$ = the 3rd angle.

    2) Make the equation: $3x + 6x + 9x = 180$

    3) Solve for x: $18x = 180$ *or* $x = 10$

    4) Plug the answer in the "Let" statement to find the 3 angles: $3(10) = 30$; $6(10) = 60$; and $9(10) = 90$

    5) Answer: The three angles measure 30°, 60°, and 90°.

Age Problems: Make a chart to show the information in the problem. This can be used as the "let" statement (or you can use it to make your written "let" statement). Make columns for the information given - now, in ten years, etc. See example below.

Example Sue is 5 years older than Ann. In 6 years, Sue's age will be 11 years less than twice Ann's age then. How old is each person now?

NAME	AGE NOW	AGE IN 6 YEARS
Ann	x	$x + 6$
Sue	$x + 5$	$(x + 5) + 6$

Steps: 1) Make an equation: In 6 years (use "age in 6 years" column). Sue's age $[(x + 5) + 6]$ will be (=) 11 years less than (−11) twice (2) Ann's age then $(x + 6)$ OR $(x + 5) + 6 = 2(x + 6) - 11$

2) Use distributive property: $x + 11 = 2x + 12 - 11$

3) Solve for x: $x + 11 = 2x + 1$

$$\underline{-x \quad -1 \quad -x -1}$$

4) Ann's current age: $10 = x$

5) Plug x back in to find Sue's age: $x + 5 = (10) + 5$, so Sue's age $= 15$

6) Answer: Ann is now 10 and Sue is now 15.

7) Check: In 6 years Ann will be 16, Sue will be 21
                                       (21 is eleven less than twice 16).

Geometry Word Problems: Draw a diagram to demonstrate the problem. Label it with the information for the let statement. If a formula is involved, like area, write the formula down. Write the given information on the diagram, then substitute what is given (what you know) in the formula. Then use that to find what you need to solve the problem. A neat, clear diagram with correct labels can be used as a "let" statement.

Example A garden has a perimeter of 48 feet. If one side of the garden is 2 feet shorter than the other side, what are the dimensions of the garden?

Diagram:

x–2

x

$P = 2(l + w)$
$48 = 2(x + x - 2)$
$48 = 2(2x - 2)$
$48 = 4x - 4$
$$\underline{+4 \qquad +4}$$
$52 = 4x$

or use a "let" statement
Let x = the width
Let $x - 2$ = the length

$\dfrac{52}{4} = \dfrac{4x}{4}$

$13 = x$ $x - 2 = 13 - 2 = 11$

Conclusion: The garden is 13 feet by 11 feet.

Integrated Algebra Made Easy

57

16 – SYSTEMS OF EQUATIONS
(Simultaneous Equations)

Linear Equation: A first degree equation which will form a straight line when its solution set is graphed. It can contain one or two variables. (See also page 90)

Linear Pairs: Two first degree equations that are solved together. Linear pairs can be solved algebraically or graphically. Their solution set as a pair of equations is the ordered pair whose values make both equations true. Sometimes a linear pair has no solution.

Graphing: Each equation will be graphed on the same coordinate axis (same graph). The solution is the point(s) of intersection of the two graphs. (See also page 75) If the lines are parallel there is no solution. If the two lines are concurrent (in the same place) the solution is all the real numbers.

Algebraic Solution: Two methods are used -- substitution or addition.(See page 90)

1) **Substitution:** One equation is manipulated so that "x" or "y" is isolated, then the resulting representation of that variable is substituted in the second original equation. The remaining variable is solved for, then that answer is substituted in either original equation to find the second variable.

Example $x - y = 1$ and $x - 2y = 3$

Steps: 1) Solve the first equation ($x - y = 1$) for "x": $\qquad x = y + 1$

2) Get the second equation ($x - 2y = 3$) $\quad (y + 1) - 2y = 3$
and substitute ($y + 1$) for "x" in it. $\qquad -y + 1 = 3$
Solve for "y": $\qquad\qquad\qquad\qquad -y = 2$
$\qquad\qquad\qquad\qquad\qquad\qquad\qquad y = -2$

3) Go back to an original equation: $\qquad x - y = 1$

4) Substitute -2 for "y": $\qquad\qquad x - (-2) = 1$

5) Solve for "x": $\qquad\qquad\qquad\quad x + 2 = 1$

6) Indicate both answers: $\qquad\qquad x = -1$ and $y = -2$

7) Checking both answers $\qquad x - y = 1 \qquad\quad x - 2y = 3$
in both original equations $\quad -1 - (-2) = 1 \qquad -1 - 2(-2) = 3$
is the final step: $\qquad\qquad -1 + 2 = 1 \qquad\quad -1 + 4 = 3$
$\qquad\qquad\qquad\qquad\qquad\qquad 1 = 1 \qquad\qquad\qquad 3 = 3$

Note: This method is recommended ONLY when the coefficient of "x" or "y" is 1. Coefficients other than 1 result in fractional substitutions which must be "cleared" or they rapidly become unmanageable.

Integrated Algebra Made Easy

Algebraic Solution continued

2) **Addition:** If two equations have the same variable with opposite coefficients, we can add the equations together and eliminate that variable. Sometimes it is necessary to multiply one equation (or both) to make equivalent equations that can be used in this method.

Steps: 1) Arrange both equations using algebraic methods so the variables are underneath each other in position.

2) The goal is to eliminate one variable by adding the two equations together. Examine the variables and their coefficients. Find the least common multiple of the coefficients of either both x's or both y's.

3) Multiply each equation by a positive or negative number so the coefficients of the variable chosen are equal and opposite in sign.

4) Add the two equations together. One variable will disappear.

5) Solve for the variable that is visible.

6) Choose one of the *original* equations and substitute the value for the known variable to find the other variable.

7) *Check* in *both original* equations.

Example Solve the following equations for x and y:
(A) $4x + 6y = 64$ (B) $2x - 3y = -28$

Steps: 1) 6 is a common multiple for 6 and 3 so leave (A) as is and multiply (B) by 2.

$$2(2x - 3y = -28) \; ; \; 4x - 6y = -56$$

2) Add (A) and (B) in order to isolate x:

Equation (A):	$4x + 6y = 64$
NEW Equation (B) from step 1:	$4x - 6y = -56$
	$8x = 8 \; ; \; x = 1$

3) Insert the answer back into either (A) or (B): $4(1) + 6y = 64$

4) Solve for y: $6y = 60 \; ; \; y = 10$

5) Check in both *original* equations.

$$4(1) + 6(10) = 64 \qquad 2(1) - 3(10) = -28$$
$$64 = 64\sqrt{} \qquad -28 = -28\sqrt{}$$

Note: Experience will help you decide which variable to work with in the addition method. Looking for variables which already have opposite signs allows you to avoid multiplying by a negative number with its associated opportunities for error. Using small multiples is helpful, too. You wouldn't want to use a common multiple for 11 and 14 if you could use the other variable and have a common multiple of 2 and 5.

WORD PROBLEMS WITH 2 VARIABLES

As in all word problems, read and reread. Make sure your equations represent the phrases in the wording of the problem.

1. Identify each unknown quantity and represent each one with a different variable in a let statement. READ CAREFULLY - make the let statement accurate.

2. Translate the verbal sentences into *two* equations.

3. Solve as a system of equations. Usually these problems are solved algebraically but follow directions - you might be directed to solve them graphically.

4. Check the answers in the words of the problem.

(See also: Word Problems page 55.)

Two Variable Word Problems: Many word problems can be set up using two variables instead of using just one. If you choose to use that method, you must then make two equations to solve. The problem will then be solved as a "system of equations."

Example Together Evan and Denise have 28 books. If Denise has four more than Evan, how many books does each person have?

Let x = the number of books Evan has

Let y = the number of books Denise has

Steps: 1) Set up equation (A): $x + y = 28$

2) Set up equation (B): $y = x + 4$

3) Use substitution (see page 55): $x + (x + 4) = 28$

$$2x + 4 = 28$$
$$\underline{-4 \quad -4}$$

4) Solve for x: $2x = 24$

$$x = 12$$
$$x + y = 28$$

5) Substitute in original: $12 + y = 28$

6) Solve for y: $\underline{-12 \quad\quad -12}$

$$y = 16$$

Answer: Evan has 12 books and Denise has 16 books.

17 – SOLVING SIMPLE INEQUALITIES

The SOLUTION SET for an inequality is a group of numbers rather than a single number. We solve simple inequalities using the same algebraic procedures that we use for equations EXCEPT when we multiply or divide by a negative number. When multiplying or dividing by a negative number, the DIRECTION of the inequality sign is reversed, < becomes >, and > becomes <.

The solution set for inequalities is not always a finite set like the answers are for equations. Write the solution set as SS = $\{x | x > 5\}$ for example. If the domain is whole numbers, then write SS = {6,7,8,...} for example. If the solution set is between certain numbers, write $\{x | 3 < x < 10\}$. If the answer is integers greater than 3 and less than 10, write SS = {4,5,6,7,8,9}. If the domain is integers or whole numbers, it may be necessary to round a fractional or decimal answer up or down so it solves the equation and is in the domain. In word problems, it is sometimes necessary to answer with a whole number answer - so read the problem carefully.

Examples

1
$$3x - 4 \le -12$$
$$\underline{+4 \quad +4}$$
$$3x \le -8$$
$$\frac{3x}{3} \le \frac{-8}{3}$$
$$x \le -2\frac{2}{3}$$

2
$$-2x + 3 > 5$$
$$\underline{-3 \quad -3}$$
$$\frac{-2x}{-2} > \frac{2}{-2} \quad \text{(reverse the > and make it <)}$$
$$x < -1$$

3 Will has $70 to spend on CD's. Each CD costs $12 including sales tax. How many CD's can Will buy?

$12x \le 70$ Use an inequality for this problem --
he has only $70 and can only spend that much or less.

$\dfrac{12x}{12} \le \dfrac{70}{12}$ Let x = the number of CD's Will can buy

$x \le 5\dfrac{5}{6}$ notice that the domain is whole numbers since he can't buy a part of a CD.

Will can buy 5 CD's.

GRAPHING SIMPLE INEQUALITIES ON A NUMBER LINE

These are simple inequalities involving one variable, not equations needing lines and shading (See page 76).

Rules: > or < : Open circle. Line with arrow goes in the appropriate direction.
≥ or ≤ : When = is included, use a closed circle.

The solution set of an inequality with one < or > is a circle (closed or open as indicated by the problem) with an arrow showing the solution set.

Example $x > -7$ $SS = \{ x \mid x > -7 \}$

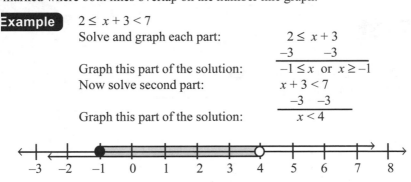

-7 -6 -5 -4 -3 -2 -1 0 1 2 3

The solution set of an inequality with two signs is where the graphs of the two separate parts overlap. Take the problem apart and graph each part. The solution set is marked where both lines overlap on the number line graph.

Example $2 \le x + 3 < 7$

Solve and graph each part:

$$2 \le x + 3$$
$$\underline{-3 \qquad -3}$$

Graph this part of the solution: $-1 \le x \text{ or } x \ge -1$

Now solve second part: $x + 3 < 7$
$$\underline{-3 \quad -3}$$

Graph this part of the solution: $x < 4$

$SS = \{x \mid -1 \le x < 4\}$ The solution set is determined by analyzing the number line where the 2 graphs have points in common. (In certain cases, there are no points in common and the solution set must be written to show this.)

To determine if a given value is a solution to an inequality, simply substitute the value given in the original inequality and see if it checks.

Example   Is 5 a solution to the inequality shown above?

$$2 \ \le \ x + 3 < 7$$
$$2 \ \le \ 5 + 3 < 7$$
$$2 \le 8 < 7 \qquad \text{Since the final result is NOT true, 5 is not a solution to the original inequality.}$$

---

18 – COORDINATE GRAPHING

DEFINITIONS AND TERMINOLOGY

Labels: Essential in graphing. Label x and y axes on the positive side of their axes (y on top and x on the right), origin, scale, line(s), and the coordinates of the solution (point of intersection) of 2 lines that are graphed if a system of equations is done. Labels are shown in the example graphed on page 75.

Linear Equation: Represented on a coordinate graph by a straight line. Linear equations have variables with exponents of "1" (not written).

❶ $y + x = 3$ ❷ $y = 2x + 4$

Coordinate Graph: Coordinate plane or grid. Each point on a coordinate graph can be located using a horizontal (x) value and a vertical (y) value called the coordinates of the point (x, y).

Origin: The intersection of the x and y-axes, the point $(0, 0)$. Label "0".

Horizontal Axis (x-axis): Arrows needed. Label in single units when possible.

Vertical Axis(y-axis): Arrows needed. Label in single units also.

Quadrant: 1/4 of the coordinate graph. The quadrants are labeled with the right upper corner of the graph as Roman Numeral I, then going counterclockwise, II, III, and IV. Quadrants do not have to be labeled unless specific instructions to do so are given.

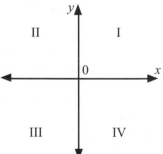

Ordered Pair: Coordinates of a point. A value for x and a value for y used to locate a point on a coordinate graph. An ordered pair is in the form (x, y). x is the horizontal value (also called the abscissa), y is the vertical value (also called the ordinate). The coordinates of a point must have a ( ) around them and be separated by a comma.

Example $(5, -3)$

Line: The line formed by an equation that is graphed on a coordinate graph represents all the values of x and y that will make the equation of that line true. The coordinates of any point on the line can be substituted for x and y, in the correct order, in the equation of the line and they will check. They "satisfy" the equation.

Collinear Points: Two or more points that are on the same line. To find out if two points are collinear, substitute the (x, y) values of one in the equation of the line. Then substitute the coordinates of the other in the same equation. If both sets satisfy the same equation, then they are collinear.

Independent Variable: This is the variable whose value is shown on the horizontal axis (x)

Dependent Variable: The variable whose value is shown on the vertical axis. (y)

Slope: The ratio of the change in y to the change in x of a graphed line – it tells the "steepness" of the graph. It is usually written as a fraction. The symbol for slope is "m" and that is used in formulas but it is not acceptable as part of the answer. The word "slope" should be used.

Slope Intercept Form: $y = mx + b$ is the Slope Intercept Form of a linear equation, where "m" represents the slope and b represents the y value of the y-intercept.

Working with Slope:

    Using Slope: Starting at a point on the line, the slope shows how to locate the next point on a graph line. The numerator (top) of the slope fraction indicates vertical movement on the graph, the denominator (bottom) number indicates horizontal movement. If slope is shown as an integer rather than a fraction, it can be made into a fraction by putting "1" in the bottom of the fraction to serve as the denominator.

Examples

    1   If the slope = 3, it becomes $\frac{3}{1}$.

         Count right 1 and up 3: $\dfrac{3 \uparrow}{1 \rightarrow}$

    Positive Slope: The line goes up as it moves to the right.

2 If the slope $= -5$, it becomes $\frac{-5}{1}$.

Count right 1 and down 5: $\frac{-5 \downarrow}{1 \rightarrow}$

Negative Slope: The line goes down as it goes to the right.

Reminder: $\frac{-5}{1} = -\frac{5}{1} = \frac{5}{-1}$. For slope, it is easiest to use $\frac{-5}{1}$

Need help remembering slope?? The "Upper" number is the "Up or" down number. The slope fraction is also alphabetical - <u>U</u>pper number is <u>U</u>p or down, or <u>V</u>ertical. <u>B</u>ottom number is <u>A</u>cross. The "Upper number" shows movement towards or away from the North Pole which is on the top of a map (which is a type of graph system, also). The axes on a coordinate graph have "y" at the top - and the y or vertical movement of slope is the top number of the slope fraction. The x label on the axis is located in a lower position on the graph - at the right edge and the x direction movement of the slope is shown in the lower position of the fraction.

Finding Slope
From 2 points:

Slope Formula: $\quad m = \dfrac{\Delta y}{\Delta x} = \dfrac{y_2 - y_1}{x_2 - x_1}$

Δ is the Greek letter delta and it stands for "change in".

To find the slope of a line if you know two points on the line, name the points 1 and 2. Then substitute the y values for 1 and 2 in the formula and the x values for 1 and 2 in the formula. These must be taken in the same order.

Example Find the slope of a line passing through (3, 2) and (–4, 5).

Steps: **1)** Point 1 is (3, 2)   Point 2 is (–4, 5)

2) Determine x and y values: $x_1 = 3$ $\qquad\quad$ $x_2 = -4$

$\qquad\qquad\qquad\qquad\qquad\qquad\quad$ $y_1 = 2$ $\qquad\quad$ $y_2 = 5$

3) Plug Points into formula: $m = \dfrac{5-2}{-4-3} \Rightarrow$ slope of the line $= \dfrac{3}{-7}$

$\qquad\qquad\qquad\qquad\qquad\qquad$ [Remember: $\dfrac{-3}{7} = -\dfrac{3}{7} = \dfrac{3}{-7}$]

From an Equation: Solve the equation for y in terms of x. Put it in the form $y = mx + b$. The coefficient of the x variable (after the equation is solved for y in terms of x) is the slope of the line of that equation.

Example   What is the slope for the line $2y - 4x = 6$?

Steps: **1)**   Solve for y:

          2)   use $y = mx + b$: $y = 2x + 3$

          3)   Determine slope (m): $m = 2$ *or* $2/1$

(The slope is used as a fraction. Use 1 for the denominator if "m" is not already a fraction.)

Using A Straight Line Already Graphed: Locate a point and then count the horizontal and vertical distance to the next point to find the slope ratio.

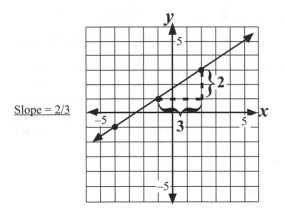

Slope = 2/3

Special Slopes - Memorize these

slope = 0

If y = a number, and the equation has no "x", the graph is a **horizontal** line and parallel to the x-axis. It crosses the y-axis at the number given in the problem. **The slope = 0.**

A line parallel to the x-axis will never crosses the x-axis. x doesn't even appear in the equation. The equation will always be $y =$ ____ (blank is filled with a number, the appropriate value of y.)

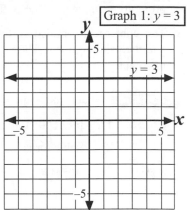

Example

Write an equation through the point (2, 3) that is parallel to the x-axis. The equation is $y = 3$.

Undefined Slope

If x = a number, and the equation has no "y", the graph is a **vertical** line and parallel to the y-axis. It crosses the x-axis at the number given and **its slope is undefined.** Sometimes we say it has no slope.

A line parallel to the y-axis will never cross the y-axis; there is no y in its equation. The equation will always be $x =$ ____ (blank is filled with a number, the appropriate value of x.)

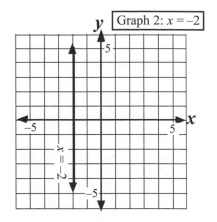

Example

Write an equation through the point (–2, 4) that is parallel to the y-axis. The equation is $x = -2$.

Y-Intercept: The point at which the graph line crosses the y-axis. Coordinates are always $(0, y)$. The x value is always 0 and y is the number on the y-axis where the line crosses.

- Solve the equation for y in terms of x and put it in the form $y = mx + b$. In this form, with y isolated, "b" is the y value of the y-intercept. As shown in the previous example, $y = 2x + 3$, the y-intercept is $(0, 3)$.

- The y-intercept can also be found by substituting "0" for x in the equation and then solving for y. The value of y when $x = 0$ is the y coordinate of the intercept.

Example What is the y-intercept for the line $2y = 4x + 2$?

Steps:	**1)**	Insert "0" for x:	$2y = 4(0) + 2$
	2)	Solve for y:	$2y = 2$; $y = 1$
	3)	Answer:	y-intercept is $(0, 1)$

Collinear points: Points that are on the same line. If the slope is the same between the first and second points as it is between the second and third (or first and third), then the points are on the same line. Use the slope formula to test this.

Example Are $(3, 5)$, $(6, 8)$, and $(-4, -8)$ collinear points?

Solution: Call $(3, 5)$ point 1, $(6, 8)$ point 2, and $(-4, -8)$ point 3.
First, find the slope between points 1 and 2.
Then use the same formula for points 2 and 3.

Slope of points 1 and 2   Slope between points 2 and 3

$$ m = \frac{8-5}{6-3} = \frac{3}{3} = 1 \qquad\qquad m = \frac{-8-8}{-4-6} = \frac{-16}{-10} = \frac{8}{5} $$

Since the slope between the first two points is 1, and the slope between the second and third points is $\frac{8}{5}$, thus not the same, these three points are *not* on the same line.

Parallel and Perpendicular Lines and Slope: Use the slope formula or the slope intercept form of an equation ($y = mx + b$) to determine if two lines are parallel, perpendicular, or neither.

- If two lines have the <u>same slope</u>, the lines are parallel or they may be the same line.

  Example Are $y = 3x - 4$ and $y = 3x + 7$ parallel lines?

  They both have a slope of 3, so yes, they are parallel lines.

  Note: If two equations have the same slope and the same y-intercept, they become the same line when they are graphed.

- If two lines are <u>parallel</u>, they have the same slope.

  Example Write an equation of a line parallel to the line $y = -3x + 2$.

  The slope of this line is -3 so the new equation will have the same slope but a different y-intercept.

  It could be $y = -3x + 5$ *or* $y = -3x - 2$, or just $y = -3x$ (the y-intercept is zero).

- If two lines have slopes that are <u>negative reciprocals</u> of each other, the lines are perpendicular.

  Example Are these two lines perpendicular? $y = 3x - 4$ and $y = \frac{-1}{3}x + 12$.

  Since the slopes are 3 and $-1/3$ which are negative reciprocals of each other, the lines are perpendicular.

- If two lines are <u>perpendicular</u>, their slopes are negative reciprocals of each other.

  Example Name two points that would be on a line that is perpendicular to the line $y = -2x$.

  To solve this, we must find two points that will work in the slope formula to make a slope of $+\frac{1}{2}$.

  An answer might be (6, 3) and (12, 6).

Finding the Y–Intercept

From an equation Use $y = mx + b$. Solve the equation for y in terms of x and put it in the form $y = mx + b$. In this form, with y isolated, "b" is the y-intercept. ["m" is the slope]

Example Find the y-intercept of $2y - 2 = 4x$

Steps: 1) Solve for y: $\dfrac{2y}{2} = \dfrac{4x}{2} + \dfrac{2}{2} = y = 2x + 1$

2) B is 1, the y-intercept is 1

Substitute zero for x The y-intercept can also be found by substituting "0" for x in the equation and then solving for y. The value of y when $x = 0$ is the y coordinate of the intercept.

Example What is the y-intercept for the line $2y = 4x + 2$?

Steps: 1) Insert "0" for x: $2y = 4(0) + 2$

2) Solve for y: $2y = 2 = y = 1$.

3) Answer: y-intercept is $(0, 1)$

From a straight line already graphed Locate the point where the line crosses the y-axis. This is the y-intercept. **_Note_:** This method is not always accurate as the point of intersection with the y-axis is not always an integer.

Writing the Equation of a line

Given the slope and one point on the line Use the Slope-Intercept form of an equation, $y = mx + b$.

- Substitute the given value of the slope for "m"
- Use the x and y values of the given point for x and y in the equation.
- Solve for "b" (the y-intercept.)
- In the equation $y = mx + b$, replace "m" with the given slope and b with the value you found in step 3.

Example Write the equation of the line through (2, 16) that has a slope of –5

Steps: 1) Slope intercept form of and equation: $y = mx + b$

2) Substitute (2, 16) for x and y; and –5 for m: $16 = (-5)(2) + b$

3) Simplify $16 = -10 + b$

4) Solve for "b": $26 = b$

5) Rewrite replacing "m" with –5 and b with +26: $y = -5x + 26$

<u>Given 2 points on the line</u>: Use the point-slope form of an equation: $(y_2 - y_1) = m(x_2 - x_1)$

- Find "m" by choosing one point to be (x_1, y_1) and the other to be (x_1, y_2) and substituting those values in the slope formula.
- Solve for "m".
- Rewrite the point-slope formula.
- Substitute the answer you got in step 1 for "m" and substitute the values of (x_1, y_1) in the equation. x_2 and y_2 will stay in the equation as letters, but the subscript "2" is removed.

Example Write the equation of the line through the points (3, 7) and (5, 15). [We'll choose (3, 7) to be point 1, and (5, 15) to be point 2.]

Steps: 1) Substitute points 1 and 2 in the equation: $(15 - 7)$ $= m(5 - 3)$

2) Solve for "m": $8 = 2m \; ; \; m = 4$

3) Rewrite the formula: $(y_2 - y_1) = m(x_2 - x_1)$

4) Substitute 4 for "m" and (3, 7) for (x_1, y_1): $(y_2 - 7) = 4(x_2 - 3)$

5) Remove the subscripts: $y - 7 = 4(x - 3)$

<u>*Note*</u>: It doesn't matter which point you choose to be point 1 or point 2. Just be consistent. Simplify this equation if instructed by doing the distributive property and collecting like terms.

<u>*Note*</u>: The point-slope equation is just the slope formula in a different form. If you treat the slope formula as a proportion and "cross multiply" you get the point-slope formula. ☺

$$\frac{m}{1} = \frac{(y_2 - y_1)}{(x_2 - x_1)} \quad \text{becomes} \quad m(x_2 - x_1) = 1(y_2 - y_1)$$

To determine if a point is on a line: This means you are basically checking to see if a point is a solution to a specified equation. Although we've been graphing the equations, it is not usually accurate enough to just look at the graph and see if it goes through the point in question. The best method is to use the original equation and substitute the *x* and *y* values in it to see if it checks.

Example Is the point (5, 7) on the graph of the line $2y = 3x - 1$

Solution: $2(7) = 3(5) - 1$ *Does this work??*

$14 = 15 - 1$

$14 = 14$ Yes!

The point (5, 7) is on the graph of the line $2y = 3x - 1$

Graphing the Line

From two points: Plot the points on the graph paper. Draw a line through both points to the edges of the paper and put arrows on both ends (Figure 1). If the problem requires a segment (like the side of a polygon), the segment should end at the given points and no arrows are needed (Figure 2).

Figure 1

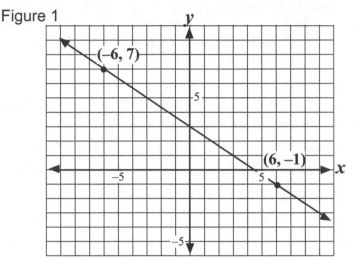

Figure 2

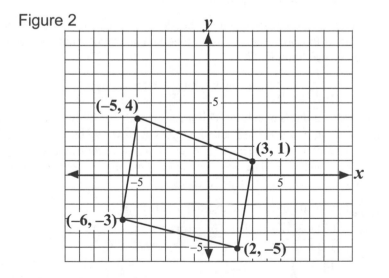

From an equation using slope - intercept: Solve for y in terms of x: $y = mx + b$

- Use "b" to find the y intercept and write it down. Locate it on the graph.
- Use "m" to find the slope and write it down.
- Start at the y-intercept point. Count to the right the number of units in the DENOMINATOR of the slope fraction (Use 1 if slope is a whole number). Then count UP or DOWN the number of units in the NUMERATOR of the slope fraction. If the fraction is positive, count UP, if it is negative, count DOWN. This locates the second point on the graph line.
- Start at the new point just located, and repeat the process in step "3".
- The 3 points should be in a straight line! If not - check your work. Draw a line (with arrows) through all three points accurately. Extend them to the edge of the graph.
- LABEL the graph line with the *original* equation.

Example Graph the equation: $3y + 2x = 9$

Steps: **1)** Isolate y: $3y = -2x + 9$

   2) Divide each side by 3: $\dfrac{3y}{3} = \dfrac{-2x+9}{3}$

   3) Solve for y: $y = \dfrac{-2}{3}x + 3$

   4) Using $y = mx + b$: $m = \dfrac{-2}{3}$

   and $b = 3$

- *Write*: "The slope is $\dfrac{-2}{3}$ and the y-intercept is $(0, 3)$."

- Plot the line: Locate $(0, 3)$ as the 1st point.

- Go to the right 3 and down 2 to find the next point.

- Repeat. Connect the points and extend the line.

- Label.

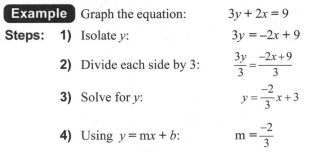
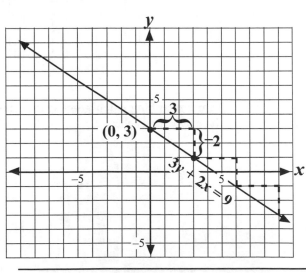

Using a Table of Values: Using the equation $3y + 2x = 9$

Solve for y in terms of x. Make a value chart. Choose three or four values of x and substitute each in the equation to find the value of y that corresponds with it. Then use those three or four sets of coordinates to plot the line. LABEL. You must show the work in evaluating the equation for "x".

x	$-\dfrac{2}{3}x + 3$	y	PLOT
3	$-\dfrac{2}{3}(3) + 3$	1	(3, 1)
0	$0 + 3$	3	(0, 3)
–6	$-\dfrac{2}{3}(-6) + 3$	7	(–6, 7)

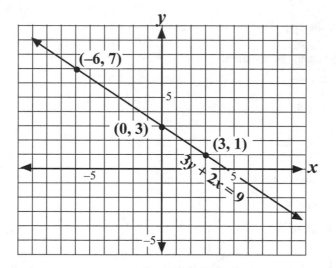

Graphing a line from a graphing calculator. Specific instructions for procedure and labels will be given by your teacher.

Graphing Systems of Linear Equations: Two or more equations are graphed on the same coordinate plane (grid).

- Graph EACH equation separately but put both on one coordinate graph. Be ACCURATE.
- Label each line as you graph it.
- The point where they intersect is the *solution set* of the system of equations.
- Label the point of intersection on the graph. This point is the solution set.
- Check both the x and y values of the solution in both original equations. The x and y values of the point of intersection must satisfy both equations.

Example Solve this system of equations graphically and check:

(A) $y = x + 4$ and

(B) $y = -2x + 1$ (These are both already in $y = mx + b$ form)

Steps: A) $\underline{y = x + 4}$ B) $\underline{y = -2x + 1}$

1) Determine the slope: slope = 1/1   slope = $-2/1$

2) Determine the y-intercept: (0, 4)   (0, 1)

3) Graph, label, and find solution set: SS = {(–1, 3)}

4) Check: (A) $y = x + 4$ (B) $y = -2x + 1$
                      $3 = -1 + 4$ $3 = -2(-1) + 1$
                      $\mathbf{3 = 3}$ √ $3 = 2 + 1 \Rightarrow \mathbf{3 = 3}$√

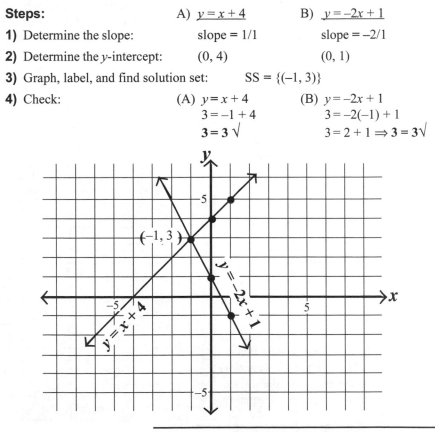

Graphing an Inequality on a Coordinate Graph: Inequalities are graphed using the same procedures that are used for graphing equations (See page 58). The line that is graphed becomes the border line of the inequality with the points on one side or the other making up the solution set. The border line itself is part of the solution set if $\leq$ or $\geq$ is involved.

1) Border lines for inequalities, broken or solid (see #3), are plotted on a coordinate graph like equation lines are. Solve each equation for y using the $y = mx + b$ form. "m" represents the slope, and b identifies the y-intercept. (Reminder: Reverse the $<$ or $>$ sign if you multiply or divide by a negative number.)

2) Plot the y-intercept point. Locate the next two points using the slope.

3) BROKEN OR SOLID? Draw the line carefully. <u>Special Rules for Inequalities</u> If the problem contains $<$ or $>$ (without an $=$ in it) the line is a broken line. If the problem includes $\geq$ or $\leq$ (includes $=$) the line is solid.

4) TEST POINT AND SHADING: Choose any point on the graph that is NOT on the line drawn. This is a test point. Put the values for (x, y) of that point in the original problem. If they make the inequality true, then shade the same side of the graphed line where the test point is. If the x and y values of the test point make the inequality false, shade the side of the line opposite the test point.

5) Label the graph: Put the original problem on the side of the graph which is shaded - place it near the graph line.

Example On a coordinate plane, graph the following inequality and label its solution set S: $2x + y < 4$

$2x + y < 4$ (*Boundary line will be broken*)

$$\frac{-2x \qquad -2x}{\quad}$$

$y < -2x + 4$

slope $= -2$

y-intercept $= 4$

Choose a Test Point: $(3, 3)$

Test it in the original:

$2x + y < 4$

$2(3) + 3 < 4$?

$9 < 4$ False

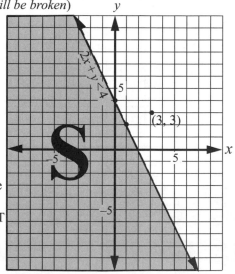

Because the test was false, shade the side of the boundary line that is opposite the side where $(3, 3)$ is located. $(3, 3)$ is NOT part of the solution set. Label the shaded part "S" and write the original inequality in that part.

Integrated Algebra Made Easy

Graphing Systems of Inequalities:

1) Graph one problem as shown on the previous page. Find the test point, shade, and label.

2) Graph the 2nd problem - same method. This time make the shading go in a different direction.

3) The section of the graph where the shading overlaps is the solution set of the inequality system. All the points in that section are in the solution set and will make both problems true. Mark a large "S" in the section that is the solution set if instructed to do so.

4) If asked to name a point in the solution set, choose any point in that overlapping area and write its coordinates in the (x, y) form. If asked to name a point not in EITHER solution set, choose a point in the section that has NO shading at all and write its coordinates in the (x, y) form.

Example Solve graphically. Label the solution set of this pair of inequalities, $y \geq x$ and $y < -x - 2$, with S. Give the coordinates of a point in the solution set.

$y \geq x$	$y < -x - 2$
Slope $= \dfrac{1}{1}$	Slope $= -\dfrac{1}{1}$
y-intercept is at $(0, 0)$ SOLID LINE*	y-intercept is at $(0, -2)$ BROKEN LINE
Test Point: $(1, 5)$ $y \geq x$	Test Point $(1, 5)$ $y < -x - 2$ $5 < -1 - 2$
$5 \geq 1$ True Shade this side of the line.	$5 < -3$ False Shade opposite side.
A point in the solution set is $(-5, 2)$	

Note: The solid line needed to graph $y \geq x$ indicates that the points on the border line are included in the solution set.

Note: INEQUALITY coordinate graphs *do require test points and shading*. Do not confuse *inequalities* with equations (See page 58).

Functions and Relations

Relation: a correspondence between two variables (x and y). When a relation is written as a set of ordered pairs, (x, y) we say that x is related to y. This set of ordered pairs is a relation. $\{(1, 5), (3, 7), (1, -3), (5, 7), (3, 10)\}$ Notice that the x values can be associated with different values of y. When $x = 3$, $y = 7$ and then in another pair, $x = 3$ is associated with $y = 10$. The $x = 3$ value has 2 different values for y.

Function: a relation in which each element of x is associated with a unique element of y. In a function, when the points are listed, there are *no* repeated values of x. Each value of x has only one value of y that is associated with it. This set of ordered pairs is a function: $\{(1, 4), (-5, 3), (4, 1), (6, 3), (-2, 7)\}$ Notice that there are no repeated x values although the values of y can be repeated. When using function notation, instead of writing $y = x$, we write $f(x) = x$.

Determining if a graph is a relation or a function: Since relations and functions are both sets of ordered pairs, we can graph them and compare them on a graph. A test called the "vertical line test" can be used to determine if a graph is a relation or a function. In this test, a vertical line (like the edge of a pencil) is moved across the graph from left to right. If the graphed line is intersected by the vertical line in more than one place at a time, the graph is a relation. If the vertical line intersects the graph in only point at a time, the graph is a function.

Examples

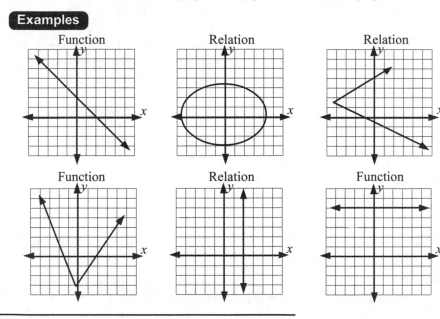

| | Function | Relation | Relation |
| | Function | Relation | Function |

Types of Graphs: There are certain shapes of graphs that are associated with various functions. The general shape of these graphs will be similar although the size and exact shape may vary based on the specific numbers used in the function. It is easy and fun to use a graphing calculator to change the size, shape, and location of the graphs of these functions by changing the coefficients of the variables and making other changes in the basic function. A few examples are shown but compare the changes using the graphing features on the calculator if possible.

Linear Function: Example: $y = x$. This is a straight line graph. The basic graph goes through the origin since the value of "b" in the familiar $y = mx + b$ form of this equation is 0. If the coefficient of x is negative, the line has a negative slope. The value of the coefficient of x determines the steepness of the graph. (If another number is added as "b" the graph moves up or down so it intersects the y-axis at another point instead of 0.)

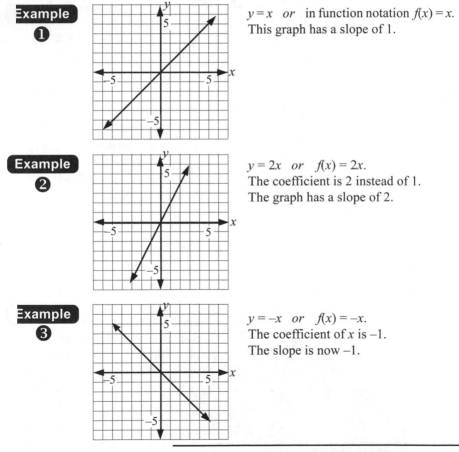

Example ❶

$y = x$ *or* in function notation $f(x) = x$.
This graph has a slope of 1.

Example ❷

$y = 2x$ *or* $f(x) = 2x$.
The coefficient is 2 instead of 1.
The graph has a slope of 2.

Example ❸

$y = -x$ *or* $f(x) = -x$.
The coefficient of x is –1.
The slope is now –1.

A quadratic (parabolic) Function: $y = x^2$ is a graph of a parabola (U shaped graph.) The parabola has its minimum point at the origin and the axis of symmetry is the y-axis. If the coefficents are changed, the graph becomes wider or more narrow. If values are added to complete the familiar $y = ax^2 + bx + c$ the graph moves right or left and up or down. If a is negative, the graph becomes "upside down" or a parabola whose opening is toward the bottom of the graph instead of the top.

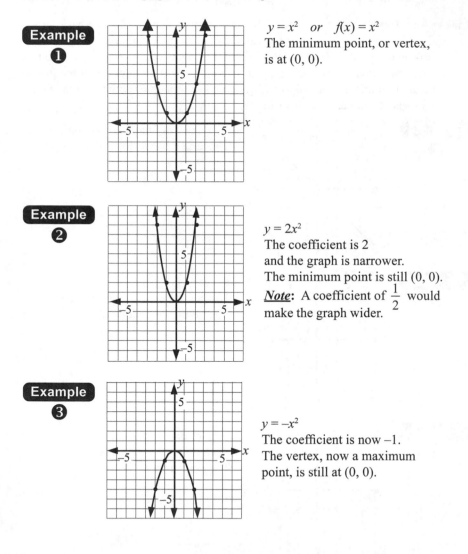

Example ❶

$y = x^2$ or $f(x) = x^2$
The minimum point, or vertex, is at (0, 0).

Example ❷

$y = 2x^2$
The coefficient is 2 and the graph is narrower.
The minimum point is still (0, 0).
Note: A coefficient of $\frac{1}{2}$ would make the graph wider.

Example ❸

$y = -x^2$
The coefficient is now –1.
The vertex, now a maximum point, is still at (0, 0).

An **absolute value function:** $y = |x|$ is a V shaped graph whose vertex is on the origin. It is symmetric to the *y*-axis. Changing the coefficient of *x* will make the V wider or more narrow. A negative coefficient OUTSIDE the absolute value symbol will make the graph an upside down V – the maximum point will be at the origin. Adding other numbers to the function will move it up or down, right or left.

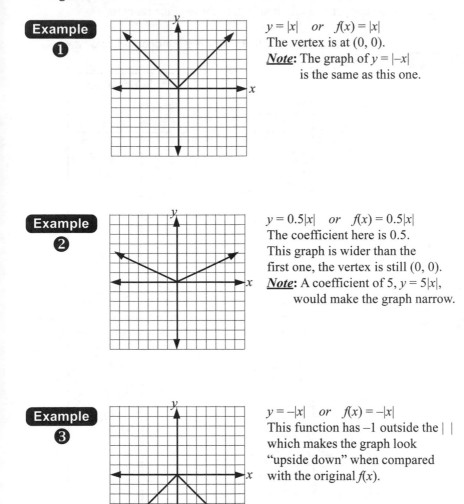

Example ①

$y = |x|$ *or* $f(x) = |x|$
The vertex is at (0, 0).
Note: The graph of $y = |-x|$
          is the same as this one.

Example ②

$y = 0.5|x|$ *or* $f(x) = 0.5|x|$
The coefficient here is 0.5.
This graph is wider than the
first one, the vertex is still (0, 0).
Note: A coefficient of 5, $y = 5|x|$,
          would make the graph narrow.

Example ③

$y = -|x|$ *or* $f(x) = -|x|$
This function has –1 outside the | |
which makes the graph look
"upside down" when compared
with the original $f(x)$.

An exponential function: $y = 2^x$. The basic graph of this function is a curve that approaches the negative x-axis, but never crosses it. The graph always crosses the y-axis at $(0, 1)$. It increases sharply as it moves to the right if x is a positive number great than 1. Changing the base (2) makes the curve even more "steep" as it rises to the right. Changing the coefficient of x to a negative number reverses the curve – it curves down moving left to right. It will then approach the positive x-axis but not cross it. It still goes through the point $(0, 1)$.

Example ❶

$y = 2^x$ or $f(x) = 2^x$. Graph approaches but does not cross the negative x-axis. The y-intercept is $(0, 1)$

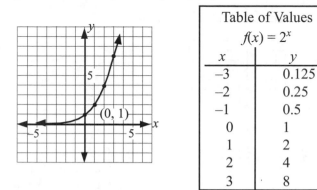

Table of Values
$f(x) = 2^x$

x	y
−3	0.125
−2	0.25
−1	0.5
0	1
1	2
2	4
3	8

Example ❷

$y = 2^{2x}$. The coefficient of x is 2. The graph increases more quickly but still intersects the y-axis at $(0, 1)$ and approaches but does not cross the negative x-axis.

Table of Values
for $y = 2^{2x}$

x	y
−3	0.01563
−2	0.0625
−1	0.25
0	1
1	4
2	16
3	64

Example
3

$y = 2^{(-x)}$ In this example, the coefficient of x is -1. The graph decreases as it goes from left to right. It has (0, 1) as its y-intercept, and now it approaches but never crosses the positive x-axis.

Table of Values for $y = 2^{2x}$	
x	y
-3	8
-2	4
-1	2
0	1
1	0.5
2	0.25
3	0.125

Note: When using your graphing calculator it is fun and easy to see what happens to the graph when function is changed in one or more ways. Compare these with the basic function graphs.

Linear: $f(x) = x + 3$; $f(x) = 3x - 5$

Quadratic: $f(x) = \frac{1}{3}x^2$; $f(x) = 2x^2 + 3x + 1$

Absolute Value: $f(x) = -2|x|$; $f(x) = |3x|$; $f(x) = |x| - 2$

Exponential: $f(x) = 2^{-2x}$; $f(x) = 2^{(x+3)}$

19 – SOLVING QUADRATIC EQUATIONS

Quadratic: A second degree equation - it contains a "squared" variable, like x^2. Quadratic equations can be solved algebraically or graphically.

Roots: Quadratic equations have two solutions which are called roots. When solving algebraically, the roots are determined by the factors. Sometimes the roots are equal. A root can be 0. Both roots will check in the original equation if they are correct. Some roots are mathematically correct but do not fit the requirements of the problem. They must be marked as rejects and it isn't necessary to check them. The roots of the quadratic equation are the points where the graph of the equation crosses the x-axis.

Solving Algebraically

1. "Move" all the terms to one side of the equal sign using algebraic methods.

2. Write the terms (usually located on the left side of the equal sign for ease in working) in standard form.

3. The other side of the equal sign is 0. (Usually the right side.)

4. If a variable with an exponent of "2" is still in the problem, you must factor. Remember to look for a common factor first, then factor the remaining equation using ( ) ( ). See Chapter 10 for review of factoring.

5. KEEP the = 0. This is an equation and must have an equal sign.

6. Separate the factors and make each set of ( ) = 0. This utilizes the zero product property - also called the multiplication property of zero. This property says that if a product equals zero, one or more of its factors must be zero.

7. Solve each equation for the variable.

   Note: The solutions or roots are the opposite numbers from those in the factors. The roots are also called the "zeros".

8. Check each answer in the original quadratic equation. Both answers must check.

9. Indicate whether both answers are to be used or if one is a reject based on the information in the problem. A reject must be clearly marked "reject.

Example Equation $$x^2 - 14 = -5x$$

Steps: **1)** Move all terms to one side of equal sign: $$x^2 + 5x - 14 = 0$$

2) Factor the equation: $$(x + 7)(x - 2) = 0$$

3) Separate the factors: $$(x + 7) = 0 \mid (x - 2) = 0$$

4) Solve both equations for x: $$\boxed{x = -7} \mid \boxed{x = 2}$$

5) Check both answers in $x^2 + 5x - 14 = 0$ $(-7)^2 + 5(-7) - 14 = 0$ √

$$49 - 35 - 14 = 0$$

6) Both check. $$(2)^2 + 5(2) - 14 = 0 \text{ √}$$

$$4 + 10 - 14 = 0$$

<u>*Note*</u>: The roots, or solutions, are the opposite numbers that are in the factors. If the equation is graphed, the graph will intersect the x-axis at -7 and at 2.

Example **of Rejects:** Negative answers for distance or age; fractions and decimals if the domain is integers; positive answers if the domain is negatives; odd numbers if the domain is evens, etc. REREAD the problem carefully before deciding if there are any rejects.

Example The length of a rectangle is 2 more than its width. The area of the rectangle is 35. Find its dimensions.

Steps: **1)** Make a diagram

2) Formula for Area $= \ell \bullet w$

3) Equation: $$(x)(x + 2) = 35$$

4) Simplify:
$$x^2 + 2x = 35$$
$$\underline{-35 \quad -35}$$
$$x^2 + 2x - 35 = 0$$

5) Factor: $$(x - 5)(x + 7) = 0$$

6) Solve: $$x - 5 = 0 \mid x + 7 = 0$$
$$\boxed{x = 5} \mid x = -7$$

7) Find the other side: $\boxed{x + 2 = 7}$ (reject -7 because

8) Answer: The rectangle is 5 by 7.   a rectangle cannot

9) Check: $(5)(7) = 35$ √   have a negative side.)

ALTERNATE ALGEBRAIC SOLUTIONS

Square Roots: When solving an equation like $x^2 - 36 = 0$, (there is no "x" term in the middle) just add "36" to both sides. Then take the square root of each side.

Example $x^2 - 36 = 0$

$$\underline{+36 \quad +36}$$
$$x^2 = 36$$
$$x = \pm\sqrt{36}$$
$$x = 6 \text{ and } -6 \quad \text{(Remember to check both answers in the original.)}$$

Completing the Square is another method of solving quadratic equations. It makes an unfactorable equation into a factorable equation which can then be solved by factoring. Completing the square is taught in a higher level math course.

Quadratic Formula: This formula is used commonly at higher levels of math to solve quadratic equations. The equation is put in standard form and the coefficients are assigned letters that are used in the formula. It is shown here for reference only.

$$x = \frac{-b \pm \sqrt{b^2 - 4ac}}{2a}$$

Quadratic Word Problems: Word problems that use the words "squared" or "the product of" often result in equations which contain a variable to the 2nd power: x^2. The let statement contains only one variable. When the equation is written, it will contain multiplication of the parts of the let statement. When the multiplication is performed, the variable will have an exponent. Quadratic word problems are usually solved algebraically but could be done graphically as well. (See next page for graphing quadratic equations.)

Example Jack had to get three more pails of water than Jill did. The product of the number of pails they had to get was 40. How many pails of water did each have to get?

Let x = Number of pails of water Jill had to get

$\therefore \ x + 3$ = the number of pails of water Jack had to get

$$x(x + 3) = 40$$
$$x^2 + 3x = 40$$
$$x^2 + 3x - 40 = 0$$
$$(x + 8)(x - 5) = 0$$
$$x + 8 = 0 \quad x - 5 = 0$$
$$\text{reject } x = -8 \quad \boxed{x = 5}$$

It isn't possible to carry -8 pails of water, so the solution -8 is rejected. Use 5 as the only correct answers for x.

Jack's pails: $x + 3 = (5) + 3$

$$\boxed{x + 3 = 8}$$

CONCLUSION: JILL HAD TO GET 5 PAILS AND JACK HAD TO GET 8.

Integrated Algebra Made Easy

Solving Quadratic Equations (functions) Graphically: Parabolas The graph of a quadratic equation in the form $y = ax^2 + bx + c$ where a, b, and c are real numbers and $a \neq 0$ is a parabola. It can also be written as a function: $f(x) = ax^2 + bx + c$. The graph has a "U" shape. The position and exact shape of the "U" are determined by using the values of a, b and c in the equation.

Vertical or Horizontal? Graphs from the equation $y = ax^2 + bx + c$ (or from $f(x) = ax^2 + bx + c$) are vertical in appearance and graphs of $x = ay^2 + by + c$ are horizontal on a coordinate plane. We will work with vertical parabolas in the examples given. Horizontal parabolas have the same basic characteristics and the "vertical" information can easily be interpreted for them although they are not functions and cannot be written using function notation, $f(x)$.

Up or Down? If "a" is negative, the graph of the parabola will open downward, if "a" is positive, the graph opens upward.

Y-Intercept: In the equation, $y = a^2 + bx + c$, the constant, "c", is the "y" intercept.

Axis of Symmetry: The line of reflection of a parabola. Points on one side of the axis of symmetry are mirror images of the points on the other side. The axis of symmetry goes through the turning point of the parabola. To find the equation of the axis of symmetry use the formula $x = -b/2a$. The axis of symmetry can also sometimes be read directly from the graph once it is completed.

Turning Point or Vertex: The maximum (if parabola opens downward) or minimum (if the parabola opens upward) point on the graph. The turning point is on the axis of symmetry. The "x" coordinate of the turning point can be found by using $x = -b/2a$. The "y" coordinate of the turning point can then be found by substituting the value of "x" into the original equation, $y = ax^2 + bx + c$. The "x" coordinate of the turning gives a "center" for the table of values used to graph a parabola. This can also be read from the graph at times.

Roots: The roots, or zeros, of the equation are the "x" values of the points where the parabola crosses the x-axis. There can be one, two, or no real roots. If it doesn't cross the x-axis at all, then there are no real roots. If it just touches the axis, there is one real root, and if it intersects it in two places, then there are two real roots. The roots can be found by reading the graph - *accuracy in graphing is essential here*. To check the roots: Since $y = 0$ at the roots, substitute 0 for "y" in the equation and substitute your answers for "x". Remember the roots are the solutions of the equation or the value of x when $y = 0$. They are the opposite numbers from the factors used to solve the equation algebraically.

Note: When solving a quadratic equation algebraically, we factor, set each factor equal to zero, and solve each one. The answers are the roots or the "zeros" of the equation. If the roots are read from a graph, an *equation of a parabola* with those roots can be developed by working backwards. Use the opposite numbers to make factors and set the product of the two factors $= y$.

Graphing A Quadratic Equation

Note: This work can be done nicely using a graphing calculator. Your teacher will advise you as to what work you will need to show on your paper when using a graphing calculator.

Steps: 1) Solve the quadratic equation for "y" in terms of "x". It will be in the form
$y = ax^2 + bx + c$,
or written in function notation as $f(x) = ax^2 + bx + c$.
Make note of the values of a, b, and c.

2) Find the axis of symmetry: $x = -b/2a$

3) Make a table of values for the parabola. Use at least three integral values on each side of the "x" coordinate of the turning point which is found by using $x = -b/2a$. (Try not to use fractions or decimals for the values of "x" you choose.) Sometimes the interval of the values for "x" to be used are given.

4) Substitute the values of "x" to find the y-coordinate of each point. SHOW SUBSTITUTIONS.

5) Plot the points and sketch the graph *accurately* and with a smooth curve.

6) LABEL the vertex and two more points on the curve – one on each side of the vertex. Label the curve with the original equation.

Example			

Steps: 1) Solve by graphing:
$y = x^2 + 4x - 5$
using the interval
$(-5 \le x \le 1)$.

x	$y = x^2 + 4x - 5$	y	(x, y)
-5	$y = (-5)^2 + 4(-5) - 5$	0	$(-5, 0)$
-4	$y = (-4)^2 + 4(-4) - 5$	-5	$(-4, -5)$
-3	$y = (-3)^2 + 4(-3) - 5$	-8	$(-3, -8)$
-2	$y = (-2)^2 + 4(-2) - 5$	-9	$(-2, -9)$
-1	$y = (-1)^2 + 4(-1) - 5$	-8	$(-1, -8)$
0	$y = (0)^2 + 4(0) - 5$	-5	$(0, -5)$
1	$y = (1)^2 + 4(1) - 5$	0	$(1, 0)$

2) Write the equation for the axis of symmetry:

$$x = \frac{-b}{2a}$$

$$x = \frac{-(4)}{2(1)}$$

$$x = \frac{-4}{2}$$

$$x = -2$$

3) Write the roots (zero's) as a solution set: SS $= \{-5, 1\}$

4) Locate its vertex (turning point) and indicate
if it is a maximum or minimum point: $(-2,-9)$ minimum pt

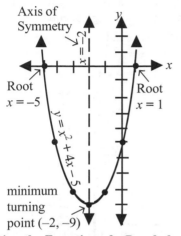

Axis of
Symmetry

Root
$x = -5$

Root
$x = 1$

$y = x^2 + 4x - 5$

minimum
turning
point $(-2, -9)$

Note: The axis of symmetry and the vertex
can both be read from the graph. The
equation for the axis of symmetry is
$x = -2$ as it is a vertical line through
the point $(-2, 0)$. The vertex on this
graph it is $(-2, -9)$ and it is a
minimum point since it is the lowest
point on the graph.

Writing the Equation of a Parabola

Equation of a Parabola: The roots of a parabola drawn on a graph can be used to
find an equation of the parabola it represents if the coefficient of the "x" term
is 1. Work backwards using factors. In the example above, the roots are -5 and
1. The factors will be formed using the numbers opposite them so the factors
will have $+5$ and -1 in them.

Steps: 1) Develop two simple equations that will result $x + 5 = 0$ $x - 1 = 0$
in -5 and 1 when solved. The numbers will be
opposite those found as the roots or solutions.

2) Working backward, make the $(x + 5)(x - 1) = 0$
equations into two factors = 0.

3) Multiply using FOIL. $x^2 + 4x - 5 = 0$

4) Since $y = 0$ at the roots, we can
exchange the 0 for "y". $x^2 + 4x - 5 = y$

Integrated Algebra Made Easy
89

20 — QUADRATIC LINEAR PAIRS

Quadratic - Linear Pair:
Two equations that are to be solved simultaneously, one is a second degree, the other is a first degree equation. In this level of math, the second degree equation would be a parabola or a circle on a graph and the first degree equation would be a straight line. There can be one, two or no solutions. The solutions are the values of x and y where the equations are equal to each other. When graphed, the two equations will cross each other at the coordinates of the solutions. The x and y values of the correct solution(s) will check in both equations.

Algebraic Solution:
Substitution is the recommended method. Solve the linear equation for "y" in terms of "x". Then substitute in the quadratic (second degree) equation.

Example Solve algebraically: $y = x^2 + 2x - 1$ and $-5x = -y + 3$

Equation A is quadratic: $y = x^2 + 2x - 1$

Equation B is linear: $-5x = -y + 3$

Steps:
1) Solve Equation B for "y" in terms of "x": $y = 5x + 3$
2) Substitute that result in Equation A: $5x + 3 = x^2 + 2x - 1$
3) Quadratic equation must be put in standard form to solve:

$$5x + 3 = x^2 + 2x - 1$$
$$\underline{5x - 3 \qquad -5x - 3}$$
$$0 = x^2 - 3x - 4$$

5) Solve by factoring: $(x - 4)(x + 1) = 0$

| $x - 4 = 0$ | $x + 1 = 0$ |

6) Two values of "x" will result: $\boxed{x = 4}$ $\boxed{x = -1}$

7) Substitute each answer for "x" in equation B to find the value of "y":
Show all substitutions.

$-5x = -y + 3$	$-5x = -y + 3$
$-5(4) = -y + 3$	$-5(-1) = -y + 3$
$-20 = -y + 3$	$5 = -y + 3$
$-y = -23$	$-y = 2$
$\boxed{y = 23}$	$\boxed{y = -2}$

8) Write the answers as ordered pairs in a solution set.
SS = $\{(4, 23), (-1, -2)\}$ or write "$x = 4$, $y = 23$ and $x = -1$, $y = -2$"

9) Check both sets of answers in both original equations.

Check (4, 23)

Equation A: $y = x^2 + 2x - 1$
$23 = (4)^2 + 2(4) - 1$
$23 = 23$ √

Equation B: $-5x = -y + 3$
$-5(4) = -(23) + 3$
$-20 = -20$ √

Check (-1, -2)

$y = x^2 + 2x - 1$
$-2 = (-1)^2 + 2(-1) - 1$
$-2 \quad = 1 - 2 - 1; \quad -2 = -2$ √

$-5x = -y + 3$
$-5(-1) = -(-2) + 3$
$5 = 5$ √

Solving Graphically: Graph and label each equation on one coordinate graph. The solution set of the system includes the point or points where the two graphs intersect each other. Use the method shown on page 88 to graph the <u>quadratic equation</u>. Then graph the linear equation using the form $y = mx + b$ where "m" is the slope and "b" is the y-intercept (or use a table of values). Label their intersections and check the solution(s) in both <u>original</u> equations.

Example Solve the following system by graphing. Then check.

$$y = x^2$$
$$y = 2x + 3$$
$$\text{slope} = m = 2$$
$$y\text{-intercept} = b = 3$$

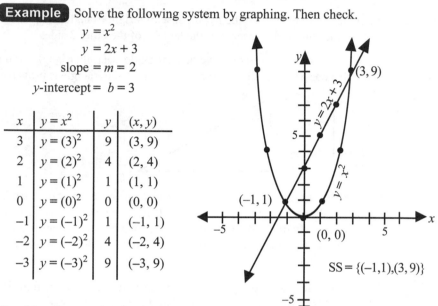

x	$y = x^2$	y	(x, y)
3	$y = (3)^2$	9	$(3, 9)$
2	$y = (2)^2$	4	$(2, 4)$
1	$y = (1)^2$	1	$(1, 1)$
0	$y = (0)^2$	0	$(0, 0)$
–1	$y = (-1)^2$	1	$(-1, 1)$
–2	$y = (-2)^2$	4	$(-2, 4)$
–3	$y = (-3)^2$	9	$(-3, 9)$

$SS = \{(-1,1),(3,9)\}$

Checking Systems of Equations

<u>**Solving Systems of Equations on Exams**</u>:
Test directions may say "solve algebraically" or "solve graphically". Do whichever method you are directed to do. Sometimes it just says "solve". In that case, the choice of graphing or using algebra is left to you. Whichever method you choose, show all work and clearly mark the answers. Check the answers in BOTH *original* equations.

<u>**Checking Systems of Equations Solved Graphically or Algebraically**</u>:
Always use the two original equations. The wrong answer will usually check in one but not in both equations. If you cannot find the error and fix it, write down "Does not check". This indicates that you know there is an error. You might receive partial credit.

21 — RIGHT TRIANGLES

The Pythagorean Theorem: In a RIGHT triangle, the side opposite the 90° angle is called the **hypotenuse** (it is always the longest side!). The other two sides are called the **legs** (they form the right angle). In the formula, the hypotenuse is labeled "c" and the legs are "a" and "b".

FORMULA: $c^2 = a^2 + b^2$

Definition: The square of the length of the hypotenuse of a right triangle is equal to the sum of the squares of the lengths of the other 2 sides.

The Pythagorean Theorem is used to prove that a triangle is a right triangle. If the measurements of the sides a triangle fit into the formula, then the triangle is proven to be a right triangle.

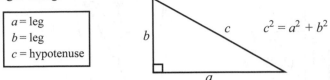

$a = $ leg
$b = $ leg
$c = $ hypotenuse

$c^2 = a^2 + b^2$

USE The Pythagorean Theorem For: finding the length of the hypotenuse of a right triangle, finding the length of a diagonal of a rectangle or square (if you know the sides), finding either of the two legs if you know one of them and the hypotenuse, finding both legs if it is an isosceles right triangle and you know the hypotenuse.

Example If the hypotenuse of a right triangle is 12 and one leg is 7, find the length of the other leg to the *nearest hundredth*.

$c^2 = a^2 + b^2$
$(12)^2 = 7^2 + b^2$
$144 = 49 + b^2$
$\underline{-49 \quad -49}$
$95 = b^2$

$a = 7$ $c = 12$ $b = ?$

$b = \sqrt{95}$ Use your calculator.
$b = 9.746...$ Write the answer shown on the calculator.
$b = 9.75$ Then round as indicated.

Pythagorean Triples: There are some common right triangle measurements (or multiples of them). Recognizing them is helpful:

Example 3, 4, and 5. (Where the legs are 3 and 4 and the hypotenuse is 5) Other common sizes of sides are; 5, 12, 13; and 8, 15, 17.

Right Triangle Trigonometry

For any right triangle, certain ratios are constant for the acute angles in the triangle. Using these ratios allows us to find a missing angle or a missing side when given some information about the right triangle. Abbreviated versions of the names of the ratios are used: Sine is Sin; Cosine is Cos; and Tangent is Tan.

Sin A is the ratio of the side opposite the given angle (A) to the hypotenuse of the right triangle. $\text{Sin} = \dfrac{\text{Opp}}{\text{Hyp}}$

Cos A is the ratio of the side adjacent to the given angle (A), to the hypotenuse of the right triangle. $\text{Cos} = \dfrac{\text{Adj}}{\text{Hyp}}$

Tan A is the ratio of the side opposite the angle (A) to the side adjacent to A.
$\text{Tan} = \dfrac{\text{Opp}}{\text{Adj}}$

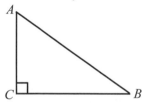

– many students use "SOHCAHTOA" to help remember the trig ratios.

Using angle A here as the given angle:

$Sin\ A = \dfrac{BC}{AB} \quad Cos\ A = \dfrac{AC}{AB} \quad Tan\ A = \dfrac{BC}{AC}$

These ratios are also expressed as decimal numbers that match specifically to an acute angle and its associated trig function. The ratios in that form can be found in a calculator or in a table of trig values.

Examples

❶ Sin 25 = 0.422618261 in a calculator.
(In a table of trig values, it would show as 0.4226.)

❷ Tan 40 = 0.839099631 in a calculator.
(In a table of trig values, it would show as 0.8391.)

To find a side of a triangle when you know one side and one acute angle:
Determine which trig function you need. Locate the angle first, then decide which 2 sides are involved in relation to the angle. Use the appropriate trig function, filling in the numbers that you know. Do the algebra with the formula and solve for the unknown side.

Examples

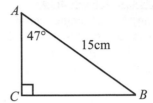

❶ Use the accompanying diagram. If A is $47°$ and $AB = 15$ cm, find the length of side BC to the *nearest hundredth*.

$Sin\ A = \dfrac{BC}{AB}$

$Sin\ 47 = \dfrac{BC}{15}$

$(15)(Sin\ 47) = BC$

$BC = 10.97030552$

$BC \approx 10.97 cm$

Note: Side AB is the hypotenuse. Side BC is opposite angle A. Use Sin A. Substitute the known numbers and do the math.

The "wiggly" equal sign ($\approx$) can be used to show that the answer is approximate – it was rounded.

❷ Using the information from Example 1, find the length of side AC. Now we are using the adjacent side to angle A and the hypotenuse. This requires Cos A.

$Cos\ A = \dfrac{AC}{AB}$

$Cos\ 47 = \dfrac{AC}{15}$

$(15)(Cos\ 47) = AC$

$AC = 10.2299754$

$AC \approx 10.23 cm$

Note: Although we did find side BC in example 1 which would allow us to use Tan 47 or the Pythagorean Theorem, it is best to use the information given directly in the problem when possible. Since the answer in example 1 was rounded, it is not completely accurate which may cause a larger margin of error if it is re-used in another part of the problem.

To find an acute angle of a triangle when given the lengths of 2 sides: Locate the angle and determine the relationship of the given sides to that angle. Set up the equation using the appropriate trig function. A decimal answer will usually be the result. This number is the actual ratio of trig function you chose for a particular angle measure. Use the "inverse" buttons on your calculator to find the angle. They are usually located directly above the sin, cos, or tan button and require the use of the 2nd button. They look like this: Sin^{-1}, Cos^{-1}, and Tan^{-1}.

Example In right triangle ABC, C is the right angle. $AC = 20$ units, $BC = 30$ units. Find the measure of angle A to the nearest degree. Draw a diagram, locate the angle and the sides involved. Decide which function is needed. In this case we have the side adjacent (AC) to angle A and the side opposite (BC) angle A. Use the Tangent function.

Step:

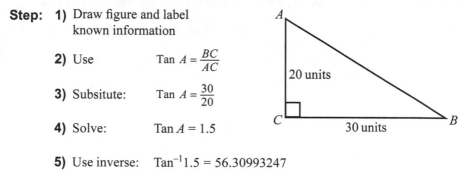

1) Draw figure and label known information

2) Use $\quad \text{Tan } A = \dfrac{BC}{AC}$

3) Subsitute: $\quad \text{Tan } A = \dfrac{30}{20}$

4) Solve: $\quad \text{Tan } A = 1.5$

5) Use inverse: $\quad \text{Tan}^{-1} 1.5 = 56.30993247$

6) Solution: $\quad A \approx 56°$

Special right triangles:
1) In a 45°- 45°- 90° right triangle (Isosceles right triangle)
   - The length of the hypotenuse equals the length of either leg multiplied by $\sqrt{2}$.
   - the length of either leg $= \dfrac{1}{2}$ the hypotenuse times $\sqrt{2}$.

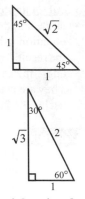

2) In a 30°- 60°- 90° right triangle
   - The shorter leg is $= \dfrac{1}{2}$ the length of the hypotenuse.
   - The longer leg is $= \dfrac{1}{2}$ the hypotenuse times $\sqrt{3}$.
   - The longer leg is equal to the shorter leg times $\sqrt{3}$.

Using "given" or known information to show or find a result using right triangles. Here are some ways to connect what you know with what you need to find.

1) Known: Measures of two sides of a right triangle.
   Find: Measure of the third side.
   - In a right triangle, the square of the hypotenuse is equal to the sum of the squares of the two legs. Pythagorean Theorem

2) Known: Measure of three sides of a triangle.
   Show: The triangle is or is not a right triangle.
   - In a triangle, if the square of one side equals the sum of the squares of the other two sides, the triangle is a right triangle. [Use this also to show the triangle is not right - if the sum of the squares of two sides does not equal the square of a third side, it is not a right triangle.]

3) Known: Lengths of two sides of a right triangle
   Find: The measure of an acute angle in the triangle.
   - Use Sin, Cos, or Tan ratios. The inverse trig button will be needed on the calculator Sin^{-1}, Cos^{-1}, or Tan^{-1} to get the angle measure.

4) Known: The measure of an acute angle and one side of a right triangle.
   Find: Another side of the triangle
   - Use Sin, Cos, or Tan ratios.

5) Known: One acute angle in a right triangle
   Find: The other acute angle
   - The acute angles in a right triangle are complementary. Remember all three angles must add up to 180°. Since one angle is 90°, the sum of the other two angles is 90°.

---

22 – QUADRILATERALS : 4 SIDED POLYGONS

Classification and characteristics of quadrilaterals can be shown like a "family tree". There are two major families at this level of math - parallelograms and trapezoids. The characteristics of any quadrilateral on the chart include all the characteristics **ABOVE** it on the chart and the ones listed with it.

Example A rhombus has all of its own characteristics and all of those of a parallelogram and all of those of a quadrilateral.

To read the chart use these rules:

1. When reading **DOWN** the chart, use the word "**some**" or "**not every**".

   Example Some quadrilaterals are parallelograms.
   Not every quadrilateral is a trapezoid, or a square, or right trapezoid.
   A parallelogram is sometimes a square.

2. When reading **UP** the chart, use the word "**all**" or "**every**".

   Example All parallelograms are quadrilaterals. Every square is a parallelogram. Squares are always rectangles.

3. When reading **ACROSS** the chart, use the word "**no**" or "**never**".

   Example No trapezoids are parallelograms.

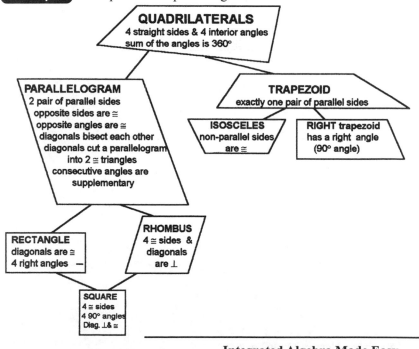

Altitude and Height of a parallelogram, a trapezoid, and also a triangle:
Parallelogram and triangle: The perpendicular line drawn from one side of a parallelogram or triangle to the vertex opposite that side is the altitude. The side the altitude is drawn from is called the base.

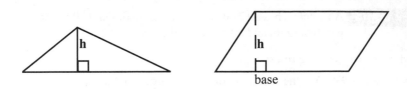

In a trapezoid, both parallel sides are often called the bases. In both figures, the height is the length of the altitude. Often the Pythagorean Theorem (See page 92). is used to find a side, or a part of the bases. The height is needed to find the area of parallelograms and trapezoids (See page 99).

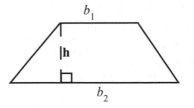

23 – PERIMETER AND AREA OF GEOMETRIC FIGURES

Note: Remember that height means the perpendicular distance from the base to the vertex opposite it. Squares and rectangles have specific formulas because the sides of the figure, being already perpendicular, are the base and height of the figure. In figures without a right angle, height is a not a side and is sometimes shown by a dotted line in a drawing. (Because the height of a figure is perpendicular to the base, it is often possible to use the Pythagorean Theorem to find the height, see page 92.)

POLYGONS

Regular Polygon: A polygon that has all sides equal and all angles equal such as a square or an equilateral triangle. If a polygon is regular and its name is not making that clear, then the word regular will be used to describe it.

Example   Find the perimeter of a regular pentagon with one side that measures 15 cm.

     Solution: Since a pentagon has five sides, and a regular pentagon has five equal sides, just multiply 5 times 15 for the answer of 65 cm.

Formulas for areas and perimeters of polygons in general:
The parts of the polygons are represented by variables in the formulas below

     l - length w - width s - side h - height b - base

     r - radius   d - diameter   A - area   P - perimeter   C - circumference

Perimeter: Distance around the sides of a polygon. Add the sides.

$$P_{rectangle} = 2l + 2w \quad or \quad 2(l + w)$$
$$P_{square} = 4s$$
$$P_{All\ others} = \text{Sum of the sides}$$

Area: Surface within the sides of a polygon. Multiply two measurements. Measured in square units or units^2.

$$A_{parallelogram} = bh$$
$$A_{rectangle} = lw$$
$$A_{square} = s^2$$
$$A_{triangle} = 1/2\ (bh)$$
$$A_{trapezoid} = 1/2\ h\ (b_1 + b_2) \quad or \quad \frac{h(b_1 + b_2)}{2}$$

Circles: $C = 2\pi r$

$A = \pi r^2$

Semi-circles - one half of a circle. $C = \pi r$ and Area $= \dfrac{\pi r^2}{2}$

Quarter Circle - 1/4 of a circle. $C = \dfrac{\pi r}{2}$ and $A = \dfrac{\pi r^2}{4}$

24 – STUDY OF SOLIDS

Solid: a closed surface, usually including its interior space.

Polyhedron: A closed figure with faces that are planar (flat). The faces meet on line segments called edges and the edges meet at vertices.

Example A closed box with flat sides.

Prism: A polyhedron with 2 congruent and parallel faces. Prisms can be square, triangular, or rectangular.

Cylinders: Solids that do not have all planar faces. A cylinder has two parallel circles for bases and a curved surface for its sides.

In this level of math, we will be working with rectangular solids (prisms) and cylinders.

Volume of Solids: Use the appropriate formula for the volume as shown below. In the formula, l is the length of the base, w is the width of the base, and h is the perpendicular measure of the solid -- from that base to the side or vertex opposite it. Radius is shown by r.

Volume of a rectangular solid = (length)(width)(height)   *or* $V = (l)(w)(h)$.

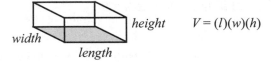
height $V = (l)(w)(h)$
width
length

When working with a cube (a rectangular solid with all 6 faces that are congruent squares), the formula is shortened to $V = e^3$ where e is the length of one edge of the cube.

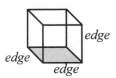
edge   All edges are equal in this cube. $V = e^3$
edge
edge

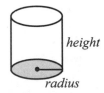
height   Volume of a cylinder = (Pi)(radius)(radius)(height)
                or $V = \pi r^2 h$
radius

---

1 Find the volume of a cube that is 3m on each side.

$V = (l)(w)(h)$

$V = 3^2 h$ *or* 3^3

$V = 3(3)(3)$

$V = 27m^3$

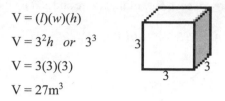

2 How much water will it take to fill a plastic container that is 12cm by 16cm by 20cm?

$V = Bh$ (Where B = area of the base)

$V = (l)(w)(h)$

$V = 12(16)(20)$

$V = 3840cm^3$

$V = 3.84$ L

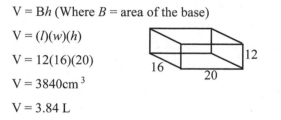

Note: 1 L = 1000cm³, so 3840cm³ is equivalent to 3.84 L

3 How many cubic inches of soup will a soup can hold if it is 5 inches high and 3 inches in diameter? Give an exact answer.

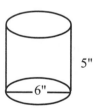

Note: They gave the diameter. The radius is needed for the formula. $r = 1.5$. Since it asks for an exact answer, we'll leave the answer in terms of π.

$V = \pi r^2 h$

$V = \pi(1.5)^2(5)$

$V = 15\pi$ *cubic inches*

If the question asked for an answer to the *nearest tenth* or *hundredth*;
- Multiply 15π in your calculator.
- Write down the entire answer.
- Round to the correct place value.

Surface Area (S.A.): This is a measure of all the outside surfaces of a solid figure. It involves finding the area of each face of the figure and adding them together.

Rectangular Prism or solid: S.A. = $2(lw) + 2(lh) + 2(wh)$

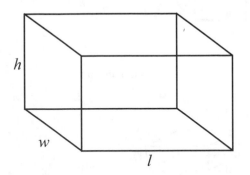

Example If a carton of paper is 15 inches by 9 inches, by 12 inches, find its surface area.

Solution: Make a diagram or assign each number a letter.
Lets make $l = 15$, $w = 12$, and $h = 9$.
Substitute the numbers in the formula.

S.A. = $2(lw) + 2(lh) + 2(wh)$
S.A. = $2 (15)(12) + 2(15)(9) + 2(12)(9)$
S.A. = $360 + 270 + 216$
S.A. = 846 square inches

Since a cube has edges that are all the same size, just use this shorter formula for a cube: S.A. = $6e^2$.

Example Find the surface area of a cube shaped box that is 4 inches on each side.

S.A. = $6e^2$
S.A. = $6 (4)(4) = 96$ square inches

Integrated Algebra Made Easy

102

Cylinder: Think of removing the top and bottom (circles) of a can and then removing the paper label. That is the surface area of the can which is a cylinder.

The areas of the two circles (top and bottom of the can) is found using the usual formula for the area of a circle. $A = \pi r^2$. Since there are two of them, we need to use it twice in the formula: $2\pi r^2$.

When the label is removed it becomes a rectangle. One side of it equals the height of the can. The other side is made from the circumference of the circular base (top or bottom). The formula for finding this area is (Circumference)(height) or $2\pi rh$

The complete formula for the surface area of a cylinder is
$$S.A. = 2\pi r^2 + 2\pi rh$$

Example Find the surface area of a cylinder with a radius of 5 ft and height of 13 ft. Give the answer in
(*a*) exact form and then
(*b*) find the answer to the nearest *tenth*.

$S.A. = 2\pi r^2 + 2\pi rh$
$S.A. = 2\pi(5)^2 + 2\pi(5)\,(13)$
$S.A. = 50\pi + 130\pi$
(*a*) $S.A. = 180\pi$ square feet
(*b*) $S.A. = 565.4866776$ which rounds to 565.5 square feet.

25 – AREA AND CIRCUMFERENCE
OF CIRCLES

Formulas for Circles:

Use: $d = 2r$ to find the diameter if you have the radius.

Use: $r = d/2$ to find the radius if you know the diameter.

Use: $C = 2\pi r$ or $C = \pi d$ to find the circumference.

Use: $A = \pi r^2$ to find the area of a circle.

Semi-Circle: 1/2 of a circle. Divide the circumference or area by 2

Quarter Circle: 1/4 of a circle. Divide the circumference or area by 4.

To find the radius or diameter if you have the circumference;

- substitute the circumference in the formula and solve for r.

- Use the same process if you have the area -substitute the value you have for area in the area formula and find the radius.

- Remember that the formula has r^2 in it, so make sure you find the square root of r^2.

- Leave π as a symbol - it will cancel in the process of solving the problem.

Pedro's family has a circular pool in their yard that has a circumference of 20π. Pedro needs to fasten a rope across the center of the pool so his little brothers can swim halfway across, rest, and then go on to the other side. He also has to order a cover for the pool for winter and needs to know the area to the nearest square foot, of the top surface of the pool.

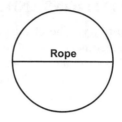

Steps:

1) Since we know the circumference is 20π, use $C = \pi d$ to find the length of the rope which is the diameter of the pool.

$$C = \pi d$$
$$20\pi = \pi d$$
$$\frac{20\pi}{\pi} = \frac{\pi d}{\pi}$$
$$d = 20$$

2) Then use $d = 2r$ to find the radius which we need to use to find the area of the surface of the pool

$$d = 2r$$
$$20 = 2r$$
$$r = 10$$

3) Use the formula $A = \pi r^2$ to find the area of the surface of the pool. Use the π button on the calculator and write down the entire calculator display of the results before rounding to the nearest square foot. The rope must be 20 feet long. The area is 100π. The winter cover has to cover 314 square feet.

$$A = \pi r^2$$
$$A = \pi (10)^2$$
$$A = 100\pi$$
$$A = 314.1592654$$
$$A = 314 \text{ sq ft}$$

Pi, π, is the ratio of the circumference of a circle to its diameter. It is an irrational number. Many times it is acceptable to leave the answer in terms of π (the symbol π will actually be in the answer). Follow the directions! You will lose credit on a problem if it says "leave in terms of π" or "find the exact answer" and you do not leave π as a symbol. π is a more exact answer than rounding. (See also—Error in Measurement —page 127)

26 – PROBABILITY

DEFINITIONS AND DESCRIPTIONS

<u>Probability</u>: The chance that a specific outcome or result will happen in an experiment. *Empirical* probability is based on the results of actual experiments that are done. *Theoretical* probability predicts the anticipated or expected result of an experiment. Probability is never less than zero or more than one.
Symbol: **P(E)**

Formula: $P(E) = \left(\dfrac{n(E)}{n(S)}\right)$ The ( ) will contain a description either in words or in symbols of the Events (successful outcomes) of the experiment.

<u>Experiment</u>: The name given to a process (like flipping a coin, or picking marbles from a bag) that is used to work with probability.

Fair or Unbiased: In predicting probability, we assume the item is "fair" and that it will not land on an edge (or on the line of a spinner between the sections). A "fair coin" is a regular, unweighted coin where there is an equal chance of heads or tails occurring when the coin is tossed. A fair spinner is a spinner with spaces of equal size for each section and a spinner that spins freely. A "fair die" is a die that is equally likely to land on any of its 6 faces.

Outcomes: The possible results from a probability experiment. The outcomes can be listed in a sample space. The total number of outcomes is **n(S)**.

Success: If an outcome is successful, that simply means it happens.

Event: A *successful outcome* of an experiment.   Symbol: **E**.
The total number of events (or *successful* outcomes) is **n(E)**.

Stage: A step or task used to perform a probability experiment. Some experiments are single stage (one step) and some are multi-stage (several steps).

<u>Independent Events</u>: Each stage of the problem occurs without regard to what happened before it.
> **Example** Toss a coin, then throw a die.

<u>Dependent Events or Conditional Probability</u>: Each stage of the problem is changed by what happened in the previous stage.
> **Example** Take a cookie from a bag, eat it, then take another cookie from the bag. The second stage has changed because one cookie has already been removed from the bag.

<u>Certainty</u>: When an particular outcome always happens, we call that a certainty. The probability of a "certainty" is 1.

Example P(H or T) on a coin toss is 1 because the coin has to land on either heads or tails.

<u>Impossibility</u>: If a particular outcome cannot happen, it is called an impossibility. The probability of an impossibility = 0.

Example P(5) on a coin toss = 0

<u>Complement</u>: The chance that a particular event will NOT occur. Symbol (~) : **P(~E)** means the probability that the event will not happen.

<u>Tree Diagrams</u>: "Groups" or sets of branches are drawn equal to the number of steps or stages to be done in the probability experiment. The first group of branches are joined on the left by a common point. Each single branch in the first group will have the next group of branches drawn off the end of it, to the right. Each branch on a tree diagram represents one or more outcomes of the experiment.

Example

Tree Diagram

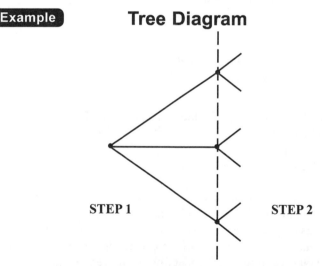

STEP 1   STEP 2

<u>Outcome Space or Sample Space</u>: A LIST of all the possible outcomes of an experiment written as ordered pairs. The ordered pairs are often listed in a rectangular arrangement. The ordered pairs can be counted to find n(S) for use in determining probability. n(S) is the number of ordered pairs or outcomes in the sample space.

Finding Probability: In order to find the probability of an outcome occurring successfully, P(E), we have to divide the number of times that a particular outcome can be successful or can happen, n(E), by the total number of outcomes that are possible in the experiment n(S).

$$\text{Use the formula } P(E) = \frac{n(E)}{n(S)}$$

Probability is usually written as a fraction, and sometimes as a % or decimal.

Examples

1 P(head) when tossing a coin is 1/2.
n(head)=1; n(S) = 2, so P(E) = 1/2

2 P(5) when tossing a regular game die is 1/6.
n(5) = 1; n(S) = 6, so P(E) = 1/6

REMEMBER: P(E) is never more than 1 or less than 0.

P(E) = 1 when the outcome is certain to happen.

Example P(a number 6 or less) when tossing a die. P($\leq$ 6) = 1

P(E) = 0 when the outcome cannot happen.

Example P(17) when tossing a die. P(17) = 0

P(~E) is the complement of P(E). To find P(~E), first find P(E), then subtract that answer from one.   P(~E) = 1 – P(E)

Example P(~5) in a die toss is 1 – P(5).
P(~5) = 1 – 1/6 = 5/6

FINDING N(E) AND N(S): In the two examples above, it is easy to find n(S) and n(E) because we are familiar with coins and dice and the numbers are small. More complicated problems may require us to use some of the following methods to find n(S) and n(E). Use one (or more than one) of these methods to count the numbers for n(E) and n(S) that you need to find the probability of an event, P(E).

1) Tree Diagram   4) Factorial
2) Sample Space or Outcome Space   5) Permutation
3) Counting Principle   6) Combination

Tree Diagrams To Find Probability

Independent Events: (Sometimes the words "with replacement" are used.)

Equal chance (Independent): Each branch represents ONE individual possible outcome in that stage of the problem like a head or a tail, a red marble or a white marble. Find n(S) by counting the number of branches at the right-hand edge of a simple tree diagram. Find n(E) by counting how many branches at the far right contain the specified outcome. Fractions showing the probability of the outcome of each branch can be placed on the diagrams although usually these simple tree diagrams are used by counting the branches that contain the desired outcome.

Example If a fair coin is tossed and a fair die is tossed, find the probabilities as indicated.

Procedure: First draw the tree diagram. It needs 2 sets or groups of branches because the experiment has 2 stages -- the coin toss and the die throw. Then count the total number of branches on the far right. That number is n(S).

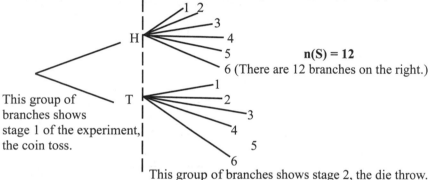

n(S) = 12
(There are 12 branches on the right.)

This group of branches shows stage 1 of the experiment, the coin toss.

This group of branches shows stage 2, the die throw.

To find **n(E)**, follow along the branches, left to right, to find the specific outcomes.

P(H,2) = 1/12 There is only one branch out of the 12 that shows the coin landing on H and the die landing on 2. **n(E) = 1**

P(H) = 6/12 Six of the branches have the coin landing on H. It doesn't matter what the die lands on in this case. **n(E) = 6**

P(3) = 2/12 Two of the branches have the die on 3. H or T doesn't matter here.

P(H,7) = 0/12 or 0. None of the branches have a 7 on them. (An impossibility.)

P (H or T) = 12/12 or 1. Every branch has either a H or T on it. The die doesn't matter here. (A certainty.) **n(E) = 12**

P(H, even number) = 3/12 Three branches have H followed by an even number.

[If fractions were shown on this tree, the coin toss branches would each be labeled 1/2 and the die throw branches would each be labeled 1/6. See page 111.]

<u>Unequal Chance Problems</u> (Independent): Each branch represents a group of items that can be outcomes. Each group of branches again represents a stage of the experiment. The experiment is an independent problem. In this type of problem, the probability of each outcome within that stage is written as a fraction on the tree diagram. Those fractions are then used to find specific probabilities. Find **n(S)** by MULTIPLYING the denominator of a fraction in stage 1 by the denominator of a fraction in stage 2.

Example

3 red marbles(R) and 2 blue marbles(B), are in a bag.

In another bag, there are 4 green pens(G) and 3 purple pens(P).

If a person picks a marble and then picks a pen,
find the probabilities indicated.

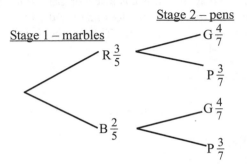

$n(S) = 5 \cdot 7 = 35$

Find **n(E)** by locating the correct branch and multiplying
the fraction in stage 1 by the fraction is stage 2.

$$P(R, G) = \frac{3}{5} \cdot \frac{4}{7} = \frac{12}{35}$$

- The $P(R) = \frac{3}{5}$ in stage 1, the $P(G) = \frac{4}{7}$ in stage 2.

- Multiply them together for the probability of choosing a red marble and a green pen.

$P(B, G) = \frac{2}{5} \cdot \frac{4}{7} = \frac{8}{35}$

$P(B, G \text{ or } P) = \frac{2}{5} \cdot \frac{4}{7} + \frac{2}{5} \cdot \frac{3}{7} = \frac{8}{35} + \frac{6}{35} = \frac{14}{35}$

$P[\sim(B, G)] = 1 - P(B,G) = 1 - \frac{8}{35} = \frac{27}{35}$ This is the complement of the $P(B, G)$.

- Use the answer you already found for $P(B,G)$ to subtract from 1.

DEPENDENT EVENTS (Sometimes called "without replacement".)

Equal Chance (Dependent): In this type of problem, the second stage depends on what happens in the first stage of the problem. The probability is handled the same way as in independent events, but the tree diagram changes.

Example There are 3 cookies in a jar--one chocolate(C), one lemon(L), and one vanilla(V). Jerry takes a cookie and passes the jar to Tom.

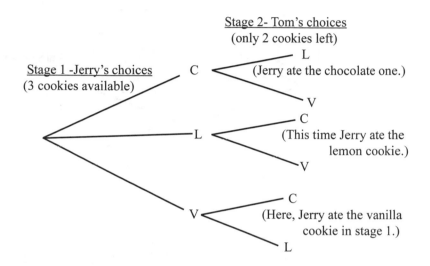

Stage 2- Tom's choices
(only 2 cookies left)

Stage 1 -Jerry's choices
(3 cookies available)

C — L (Jerry ate the chocolate one.) — V

L — C (This time Jerry ate the lemon cookie.) — V

V — C (Here, Jerry ate the vanilla cookie in stage 1.) — L

n(S) = 6 because there are 6 branches at the far right and this tree does not need fractions. Each branch represents an equal chance.

$P(C, L) = \frac{1}{6}$. Only one branch has chocolate followed by lemon.

$P(C, L \text{ or } L, C) = \frac{2}{6}$. Add the probabilities when "or" is used.

$P(C) = \frac{4}{6}$. The top 2 branches both have chocolate in stage 1, and then 2 more branches have chocolate in stage 2.

Unequal Chance (Dependent): In these problems, there is not an equal chance for items to be chosen. In stage 1, an item is chosen and not put back. Then in stage 2 an item is chosen. NOTICE THE CHANGE IN THE FRACTIONS. The probability works the same as it did in unequal chance independent on page 110.

Example A cooler contains 3 colas(C) and 1 lemon soda(L).
Brianne takes out a drink and then Ian takes one.

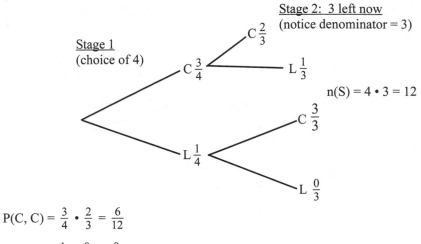

Stage 2: 3 left now (notice denominator = 3)

Stage 1 (choice of 4)

$n(S) = 4 \cdot 3 = 12$

$P(C, C) = \frac{3}{4} \cdot \frac{2}{3} = \frac{6}{12}$

$P(L, L) = \frac{1}{4} \cdot \frac{0}{3} = \frac{0}{12} = 0$

$P(\text{Ian gets C}) = P(C, C) + P(L, C) = (\frac{3}{4} \cdot \frac{2}{3}) + (\frac{1}{4} \cdot \frac{3}{3}) = \frac{6}{12} + \frac{3}{12} = \frac{9}{12}$

$P(\text{Ian gets L}) = P(\text{he doesn't get cola}) = 1 - \frac{9}{12} = \frac{3}{12}$

FINDING P(E) IN MULTI-STAGE PROBLEMS INVOLVING "AND" OR "OR"

AND: $P(A \text{ and } B) = P(A) \cdot P(A)$. The probability that event A **and** B can occur is found by multiplying the probability of A occurring by the probability of B occurring. $P(A \text{ and } B) = P(A) \cdot P(B)$. "**And**" can be indicated in probability by using the word, and by using the symbol "∧", or by writing the problem with a comma in it.

Example $P(1, 5)$ means the probability of getting 1 on the first step, **and** then getting 5 on the second step.

To use a tree diagram in a problem with "and", move across the tree from left to right following the branches that fit the problem conditions. Multiply the fractions along that path across the tree.

OR: $P(A \ or \ B) = P(A) + P(B) - P(A \ and \ B)$. The probability that event A *or* B can occur is found by adding the probability that A occurs to the Probability that B occurs, then subtracting any overlap. "**Or**" is used as a word or as symbol "$\vee$".

Example In a deck of cards, P(red card or king) $= \dfrac{26}{52} + \dfrac{4}{52} - \dfrac{2}{52} = \dfrac{28}{52}$

There are 26 red cards in a deck and 4 kings in a deck.

Since there are 2 red kings, we have to subtract $\dfrac{2}{52}$ to avoid counting them twice.

To use a tree diagram with "or" problems, read the tree vertically. The fractions are added when using "or".

It isn't possible to show all the kinds of "and" and "or" probability questions that might be asked. **Each problem must be read carefully and the probability answers determined individually** - sometimes by using both "and" and "or".

SAMPLE SPACE or OUTCOME SPACE: A display of all the possible outcomes of an experiment listed as ordered pairs. This method is usually used only in fairly small problems.

Example If a coin is tossed and then a die is thrown, use a sample space to find the probabilities indicated.

(H, 1)	(H, 2)	(H, 3)	(H, 4)	(H, 5)	(H, 6)
(T, 1)	(T, 2)	(T, 3)	(T, 4)	(T, 5)	(T, 6)

Steps:

1) Count the items to find $n(S)$ which is 12.

2) Then count the items that match the specifics for the P to find $n(E)$.

3) $P(H,1) = \dfrac{1}{12}$

4) $P(H, \text{even} \#) = \dfrac{3}{12}$

5) $P(H \text{ or even } \#) = P(H) + P(\text{Even} \#) - P(\text{both together})$

6) $\dfrac{6}{12} + \dfrac{6}{12} - \dfrac{3}{12} = \dfrac{9}{12}$

Counting Methods

Many types of problems require that we count a number of arrangements, groups, etc. There are also several ways we can figure out how many outcomes are possible for a specific set of circumstances in an "experiment." These methods can be used in probability to find $n(S)$ and/or $n(E)$ as well as in other types of problems not specifically involving probability. Formulas are used in this section for factorials, permutations, and combinations (optional) although *most calculators have these formulas* built in. Be careful when you are doing complex problems -- the calculator can't do it all. *You have to get the set-up right before you can calculate anything.*

Counting Principle: In experiments with two or more steps or stages, the total number of outcomes that can occur can be found by using the counting principle. Multiply the number of ways step 1 can occur by the number of ways step 2 can occur, step 3, etc. This equals the total number of ways in which both activities can be performed or $n(S)$ - the total number of outcomes in the sample space. This method is used when problems are very large or complex.

> **Example** Pick an ace from a deck of cards, then roll a 6 with a die.
> The total outcomes for this experiment are 52 (# of cards) times 6 (faces on die) which is 312. $n(S) = 312$.
> Probability of drawing an ace followed by
> throwing a six is: $\dfrac{4}{52} \cdot \dfrac{1}{6} = \dfrac{4}{312}$.

It would be difficult to make a tree diagram or a sample space to illustrate this problem!

The **Counting Principle** is also used for problems where probability is not the main focus — The question may ask "How many ways can something happen?" Multiply the number of ways the first part can happen by the number of ways the 2nd part can happen, etc.

> **Example** How many outfits can be made from 5 shirts and 3 pair of pants?
> $(5)(3) = 15$ outfits.

Factorials ($n!$)

A Factorial is written mathematically as $n!$ and is equal to $(n)(n-1)(n-2)(n-3)...(1)$ Special note: $0! = 1$

> **Examples** ❶ $5! \Rightarrow 5 \times 4 \times 3 \times 2 \times 1 \Rightarrow 120$
>
> ❷ Evaluate **6!** means to multiply $6 \times 5 \times 4 \times 3 \times 2 \times 1 = 720$

Multiplying and dividing factorials: Each factorial must be written showing its factors separately, then all the factors are multiplied or divided by each other.

> **Examples**
>
> ❶ $3! \cdot 4! = (3 \cdot 2 \cdot 1)(4 \cdot 3 \cdot 2 \cdot 1) = 144$ Do each **n!** separately.
>
>   (*Do not* multiply $3 \cdot 4$ first to get 12! which is $= 12 \cdot 11 \cdot 10 \cdot 9 ... 1$
>   This would create a huge number and the *wrong* answer!)
>
> ❷ $\dfrac{5!}{3!} = \dfrac{5 \cdot 4 \cdot 3 \cdot 2 \cdot 1}{3 \cdot 2 \cdot 1} = \dfrac{5 \cdot 4 \cdot \cancel{3} \cdot \cancel{2} \cdot \cancel{1}}{\cancel{3} \cdot \cancel{2} \cdot \cancel{1}} = 5 \cdot 4 = 20$
>
>   (Notice that $3 \cdot 2 \cdot 1$ cancels.)

Factorials can be used in some multistage probability problems to determine the number of outcomes in the sample space $n(S)$ which then becomes the denominator of the probability fraction.

Example What is the probability that the letters "MATH" will be drawn in that order if 4 scrabble tiles A, M, H, and T are put in a bag and drawn out at random, one at a time, and are not replaced.

Solution: Find the total possible outcomes of words that can be made with the 4 letters $n(S)$. When a letter is used it is not available for the next position. Then find P(MATH).

Think (or write down!)   <u>4 letters available,</u>   <u>then 3,</u>   <u>then 2,</u>   <u>then only 1 is left</u>
4 places for letters.   Position 1   Pos. 2   Pos. 3   Position 4

$n(S)$ or total outcomes would be $4! = 4 \cdot 3 \cdot 2 \cdot 1 = 24$. The only way the word MATH can be made is by drawing exactly M the first time, exactly A the second, and so on. So this is $1 \cdot 1 \cdot 1 \cdot 1 = 1$.   P(MATH) $= 1/24$

Permutations: Use factorials to find arrangements of elements. The example above is a permutation of 4 letters. Sometimes not all factors are used.
(See examples on page 116)

Symbols: $_4P_4 = 4 \cdot 3 \cdot 2 \cdot 1 = 24$.   It is read "A permutation of 4 things
taken 4 at a time." This is the kind of permutation used for the
letters A T M H as shown above.

If the problem was to find out how different arrangements of 4 letters could be made from all the letters of the alphabet (26 letters) without using the same letter more than once, we would write $_{26}P_4$.

This equals $26 \cdot 25 \cdot 24 \cdot 23 = 358,800$.

The probability of getting MATH $(1 \cdot 1 \cdot 1 \cdot 1)$ out of *all 26 letters* is $1/358,800$.

Additional problems encountered using factorials and permutations: Special circumstances about an arrangement are sometimes given. Each problem must be analyzed carefully. When used as part of a probability problem, these do not lend themselves to tree diagrams or sample spaces. They require individual analysis.

Examples

❶ *a)* How many 4 letter arrangements can be made of the letters A, C, E, D, M, N, T, S if the first letter must be M?

b) What is the Probability of "MATH" being one of the arrangements found in part (a) ?

Answer: *a)* 1st position must be M, so 7 of the 8 letters are left for 2nd position, 6 for 3rd, 5 for 4th. This can be written $_1P_1 \cdot {_7P_3}$ which is $1 \cdot 7 \cdot 6 \cdot 5 = 210$ arrangements.

b) BE CAREFUL of P(MATH) here! P(MATH) = 0
[there is no H available!]

❷ If a female must be Treasurer, then how many arrangements of 5 boys and 6 girls can then be chosen to be President, Vice President, and Secretary, in that order? (In problems about committees where a certain person must be President, or Vice President, make the special circumstance position the first position filled, then figure out what is left for each of the other positions and multiply.)

Answer: Treasurer could be any of the 6 *girls*. That leaves 10 *students* for President, 9 for Vice President, and 8 for Secretary.

Use the simple counting principle here.
$6 \cdot 10 \cdot 9 \cdot 8 = 4320$ different arrangements.

Note: 6 girls could fill the office of Treasurer, but, of course, only 1 girl is elected. The other 5 girls along with the 5 boys are available for the offices of President, Vice President and Secretary. This makes 10 people available for the second office filled., 9 for the 3rd, and 8 for the 4th. It is easiest to just use the simple counting principle.

Suppose the Vice President and Treasurer must both be girls?
and the Secretary and President must be boys?

Then: $6(VP) \cdot 5(Tr.) \cdot 5(Pres.) \cdot 4(Sec.) = 600$

Permutations with Repetition: In the previous examples, the items in the arrangement were all different. There are times when the some of the items to be arranged are identical to each other. The formula for this is $\frac{n!}{r!}$ where "n" is the number of items to be arranged and "r" is the number of items that are identical. (Notice that "r" is used for a different quantity in this formula than in a non-repeating permutation. Here it is used for "repeats". In the previous formula, "r" indicated how many of the available items were used at a time.)

Example How many ways can the letters in "BABIES" be arranged if all the letters are used in each arrangement? "n" is 6 and since there are two "B"s, "r" is 2. The numbers to be substituted in the formula are 6! for "n" and 2! for "r".

$$\frac{6 \cdot 5 \cdot 4 \cdot 3 \cdot 2 \cdot 1}{2 \cdot 1} = 360$$

Multiple "repeats" of letters like the letters in "**TENNESSEE**" are handled in this way: The number of different arrangements of the letters in "**TENNESSEE**" requires us to use $n = 9$. **E** appears 4 times, **N** appears twice, and **S** appears twice.

The formula will be $\dfrac{9!}{4! \cdot 2! \cdot 2!} = \dfrac{9 \cdot 8 \cdot 7 \cdot 6 \cdot 5 \cdot 4 \cdot 3 \cdot 2 \cdot 1}{4 \cdot 3 \cdot 2 \cdot 1 \cdot 2 \cdot 1 \cdot 2 \cdot 1} = 3780$

(The numerator of the formula remains the total number of items available for the arrangement.)

REMINDER: Capital P is used for Probability and for Permutations. $P(\ )$ indicates probability and $_nP_r$ indicates a permutation. Since permutations are sometimes used to find $n(\mathbf{E})$ and $n(\mathbf{S})$ in the probability formulas, be careful about which P is being (or needs to be) used! Also - watch out for the "r's" in the permutation work. Sometimes "r" stands for repeats, sometimes it stands for how many items are used at a time.

Here is a fairly complex counting problem. It requires careful analysis. Enjoy!
The owner of a baseball team insists that he make the decisions about how the team should be lined up to bat. Each day he tells the general manager how to do the line-up.

On Saturday the owner said to just "line them up". There are 9 players, and 9 positions batting, so $_9P_9 = 9! = 362,880$. This is the number of ways the team can be lined up to bat. (It's a good thing it was a night game because it took all day for the computer to print out all the lists!)

On Sunday, the owner said he wanted the pitcher to bat last. So the last position must be filled by only 1 person, the other 8 positions can be filled in any order by the remaining 8 players. $_8P_8 \cdot 1 = 8! \cdot 1 = 40,320$ lineups!

For Tuesday's game, the owner stated that either the pitcher or the shortstop had to bat last. The last position *could* be filled by 2 men, but only 1 would *actually be* batting last. That would leave 8 for the rest of the batting lineup. This time, $_8P_8$ or 8! has to be multiplied by $_2P_2$ or 2! because there are 2 ways to fill the last batting position. This gives the manager 80,640 ways to line up the batters!

Wednesday, the pitcher had to be last and short stop first, so we would have $1 \cdot 7! \cdot 1$ The pitcher and the shortstop are both assigned to their batting positions first, leaving 7 players to be arranged to bat in the middle of the lineup. This gives the manager only 5,040 ways for the lineup to be arranged.

On Thursday, the manager quit -- he hates math and was afraid to hear how many lineups the owner would demand for Friday's game ! ☺

A general rule with this type of problem is to fulfill the specific requirements of the problem first, then work with the items (in this case ballplayers) that have not already been assigned. READ CAREFULLY!

Combinations: A "combination" allows us to compute the number of ways a group can be made *without regard to order*. A very familiar example is the choosing of a committee to run a school dance. It doesn't matter who is chosen first, second, etc. The symbol used for a combination of "n" things taken "r" at a time is $_nC_r$. If "n" and "r" are equal, then $_nC_r = 1$. This is reasonable because you have six students to choose from to make a committee of six people, there is only one possible combination of students that can be on that committee! If $r = 0$, then $_nC_0 = 1$ since there is only one way a choice of zero people can be made.

The Combination Formula uses permutations in its computation. It is stated "To find the number of combinations of "n" things taken "r" at a time, divide the number of permutations of "n" things taken "r" at a time by "$r!$" and is written in symbols:

$$_nC_r = \frac{_nP_r}{r!} \text{ and since } _nP_r = \frac{n!}{(n-r)!}, \; _nC_r \text{ can be written } \frac{n!}{r!(n-r)!}$$

Examples

 If a deck of 52 cards is dealt to 4 people, each person will receive a hand containing 13 cards. How many different hands of 13 cards each can be made with a deck of 52 cards?

Analysis: Since order is not important here, this is a combination problem.

$$_{52}C_{13} = \frac{52!}{13!(52-13)!} = 635,013,559,600$$

 If a committee of three is to be chosen from a class of 15 students, how many different committees can be formed?

Analysis: Order is not important so this is a combination.

$$_{15}C_3 = \frac{_{15}P_3}{3!} = \frac{15 \bullet 14 \bullet 13}{3 \bullet 2 \bullet 1} = 455$$

In the above example, if Amy, Sue, and Joe are chosen for the committee, it doesn't matter what order they are placed in. They are simply the 3 committee members. Another example is shown below that has different circumstances.

❸ If 15 students were running for the offices of President, Vice-President and Treasurer of a club, how ways can the officers be chosen?

Analysis: Order would matter so this is a permutation problem. If Sue, Joe and Amy were the officers, Sue could be President, Joe could be V.P. and Amy could be Treasurer. The same students could be officers in a different order - Amy as Pres., Joe as V.P. and Sue as Treasurer, etc. Since the *placement* of each student as an officer makes a difference, we must use $_{15}P_3 = 15 \bullet 14 \bullet 13 = 2730$. There could be 2730 different sets of officers for this club out of the total of 15 students.

❹ Using the same 15 students, suppose Amy or Joe had to be President and the remaining 14 students were to be formed into committees of 2 each. How many ways can the group be organized? This would be 2! for the position of President -- Joe or Amy, and $_{14}C_2$ to form the committees from the remaining 14 students.

$2! \bullet _{14}C_2 = 2(91) = 182$ ways to organize the students.

27 – STATISTICS AND
REAL WORLD APPLICATIONS

Statistics is the mathematics of collecting, organizing, and analyzing data. It is an essential part of everyday life.

COLLECTING DATA

Qualitative Variable: A variable that does not have a numerical value.

> **Example** Eye color, kind of pet owned

Quantitative Variable: A variable that measures or counts - it has meaning as numbers.

> **Example** Grades on a test, hours watching TV

Single (ungrouped) statistics: Small samples or collections of data may be treated as single items. Statistical information can be obtained by working with the items of data listed singly.

> **Example** test scores of 42, 74, 75, 78, 78, and 80.

Grouped Statistics: When large numbers of items are included, the data is often treated in groups called intervals. Statistical measures are then taken from the frequencies in the intervals of the data.

Univariate Data: Measurements are made on only one variable per observation.

> **Example** Ages of the students in a club.

Bivariate Data: Measurements are made on two variables per observation.

> **Example** Grade level and age of the students in a school.

Biased: Data that is obtained that is likely to be influenced by something - giving a "slant" to the results.

> **Example** Asking people having lunch in the cafeteria of a car factory what kind of car they own. The chances are high that most of them own the kind of car made by that company.

> **Example** Standing outside Yankee Stadium and asking people coming out of a game to name their favorite team. Most of them would say.... Yankees!

Unbiased: Data that is obtained that has no connection with anything that would influence the outcome.

> **Example** Asking people leaving a large grocery store what their favorite brand of soda is. Or asking people coming out of a stadium what kind of pets they have.

ORGANIZING THE DATA

Tally Sheet: A chart showing intervals that is used to count the number of items in each interval. The statistician (that's you!) puts a mark in the appropriate interval for each item of data, crossing the marks every five for easy counting.

Frequency Table: A chart showing the intervals and the frequencies of each interval. This is often combined with a tally sheet.

Graphs: Frequency histograms, cumulative frequency histograms, whisker plots, scatter plots, and other types of diagrams that demonstrate the data.

ANALYZING THE DATA

Frequency: The number of items of data classified in any interval. The frequencies are found after classifying the data in a tally chart. Frequency is plotted on the vertical axis of a histogram.

Cumulative Frequency: The sum of the frequencies of the intervals "at or below" or "at or above" a given interval. The accumulation generally begins in the interval nearest zero if nothing else is indicated by the problem.

Intervals: The data is classified in EQUAL segments of the approximate range. Intervals are designed to fit the particular set of data. The interval on the frequency histogram must be shown as EQUAL portions on the horizontal axis of the graph.

Range: Difference between the highest and lowest number in the data.

> **Example** Grades that are all between 80 and 42. Range = 80 – 42 = 38.

Mean: Average. Add: $42 + 74 + 75 + 78 + 78 + 80 = 427$
Then divide by the number of data: 6 $427 \div 6 = 71.17$
(Note - see "outliers" Page 122.)

Median: The middle item when the data is listed in order, smallest to largest, or largest to smallest. Find the number, n, of items of data and divide by two to locate the median. If "n" is even, find the numerical average of the middle two items of data. In the above example, "n" is 6, so the median is the numeral average of the third and fourth numbers: $(75 + 78) \div 2 = 76.5$.

Mode: The measure that occurs most frequently. There can be more than one, there also can be none. In our example, 78 is the mode. Use the Mode to analyze the central tendency with qualitative data - data that is not numerical.

Percentiles and Quartiles: These measures show how a particular item of data compares to the other pieces of the data.

Example If a person's height is in the 90th percentile for his age, that means that approximately 90% of the students who were measured were shorter than she/he is.

Bias in Graphing: It is important when analyzing a graph to make sure the graph itself is accurate in its presentation of the data. Two ways a graph can be biased are if the scales are not appropriate and if there is a break in the presentation of the data.

Outliers: A data point (or points) found far outside most of the rest of the points in the data set. Outliers can strongly affect the mean. Where outliers exist, the mean is not a good representation of the center of the data.

Cumulative Frequency Table: A chart showing the intervals and the cumulative frequencies for each interval.

Histogram: USE GRAPH PAPER AND A STRAIGHT EDGE. A graph similar to a bar graph which shows the frequency in each interval of a particular set of data. There are no spaces between the bars on a histogram. The vertical axis is used for frequency, the horizontal axis shows intervals. Both axes must be labeled in equal portions. (Each interval must be the same width, each count on the frequency must be the same height.) The horizontal axis of intervals should have a spacer break, ─/\/between zero and the first interval UNLESS the interval actually starts at zero. The first interval contains the numbers closest to zero. THE LABELS OF FREQUENCY AND INTERVALS ON THE HISTOGRAM MUST MATCH THE TABLE YOU ARE USING.

Frequency Histogram: Intervals must match the frequency table and be equal portions of the horizontal axis which must be labeled "intervals". Frequency is shown on the vertical axis and must be labeled "Frequency". The title of this histogram should include "Frequency Histogram" with a reference to the data.

Cumulative Frequency Histogram: The bars represent the sum of the frequencies up through a given interval. Start with the interval nearest zero. The last bar is the highest and shows the total frequency. LABELS must match the cumulative frequency table with the problem. Vertical axis must be labeled "Cumulative frequency". The title should include "Cumulative Frequency Histogram" with a reference to the type of data included.

STATISTICAL MEASURES USING GROUPED DATA:

Mode: The mode indicates the item that appears most often. In grouped statistics, it is often referred to as the "model interval" and is located by examining a frequency table and finding the interval with the highest frequency. Sometimes two intervals have equally high numbers of frequencies and the study is said to be "bi-model", There also can be no mode. On a frequency histogram, the model interval is located by finding the highest bar.

Mean: The average of the data. For this level of statistics, using the "add the items, divide by the number of items" method is appropriate. (In later courses, formulas will be used with intervals and frequencies to find the mean as well as other statistical measures.) The interval of the mean is the interval that contains the mean.

Median: The interval of the median is the interval that contains the middle item of data – (or the average of the middle two items of data when there is an even number of measures in the data).

Percentiles: A measure at or below which a certain % of the measures fall.

Quartiles: The total frequencies are broken into 4 equal parts.
Lower Quartile - 25th percentile (.25 is the decimal equivalent).
Middle Quartile - 50th percentile or median is 50% of the total frequency (.50).
Upper Quartile - 75th percentile (.75).

FINDING PERCENTILES/QUARTILES:
2 methods:
A) Use Frequency Table (or Cumulative Frequency Table).
1. Multiply total frequency by the decimal equivalent of the percentile.
2. Locate the position of that item of data in the intervals on the table.
3. Answer carefully – usually the interval labels from the frequency table are used.

B) Use the Cumulative Frequency Histogram.
1. Multiply the total frequencies by the decimal equivalent of the percentile.
2. On the vertical axis of the cumulative frequency histogram, mark the number found in step 1.
3. Draw a horizontal line from that number to the bars on the graph.
4. The first bar that the horizontal line strikes is the interval containing that percentile.
5. The answer from the cumulative frequency table may have to be rewritten to match the labels on the frequency table - read the question carefully.

Sample Problem: A set of test grades were:

Example 82, 84, 84, 75, 78, 68, 67, 98, 87, 86,
86, 92, 68, 87, 89, 75, 66, 77, 89, 90.

Complete a tally sheet, frequency table, and cumulative frequency table. Make a frequency histogram and a cumulative frequency histogram. Find the interval containing the mean, median, and mode of the data. Locate the intervals containing the lower and upper quartiles, and the 95th percentile.

FREQUENCY TABLE

Interval	Tally	Frequency
96 –100	l	1
91 – 95	l	1
86 – 90	ⅢⅡ	7
81 – 85	Ⅲ	3
76 – 80	Ⅱ	2
71 – 75	Ⅱ	2
66 – 70	ⅢⅠ	4

CUMULATIVE FREQUENCY TABLE

Interval	Cumulative Frequency
66 – 100	20
66 – 95	19
66 – 90	18
66 – 85	11
66 – 80	8
66 – 75	6
66 – 70	4

FREQUENCY HISTOGRAM
Grades

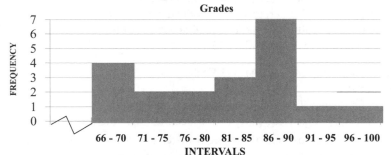

CUMULATIVE FREQUENCY HISTOGRAM
Grades

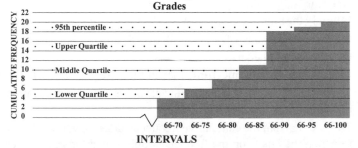

Integrated Algebra Made Easy

Mean: Add all the numbers listed in the example on page 124, then divide that total by the number of items (20): $1628 \div 20 =$ is 81.4. 81.4 is in the interval 81-85.

Median: List the numbers in order, smallest to largest. There are 20 numbers so the median is the average of the tenth and eleventh items. Since both are 84, the median is also 84. The interval is 81-85.

Mode: Look at the Frequency Histogram on page 124 - the tallest bar is the model interval. On the frequency table, it is the interval with the highest frequency. The model is in the interval 86-90.

QUARTILES AND PERCENTILES

Lower: (.25)(20) = 5   The 5th item is in the 71-75 interval.

Middle: (.50)(20) = 10 The 10th item is in 81-85 interval.

Upper: (.75)(20) = 15   The 15th, item is in the 86-90 interval.

95th percentile: (.95)(20) = 19   The 95th percentile is at 19 which is in the interval 91-95.

Box and Whisker Plot: A graph that shows how data is distributed.

To make a Box and Whisker Plot using the data from page 124:

- Plot the median (middle) of the data (84).

- Plot the first quartile (75) and the third quartile (87) (upper and lower 25% of the data).

- The "boxes" are determined by the median and quartiles.

- The "whiskers" are determined by the high and low data values.

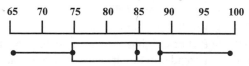

Scatter Plot – **(Scattergram):** A graph used to show a relationship between bivariate (two sets) data.

TABLE:

Test Grade	Study Time
82	2.5
84	3
84	3
75	1
78	1
68	1
67	1.5
98	5
87	4
86	4
86	3
92	4.5
68	1
87	5
89	4
75	1
66	3
77	3
89	5
90	5

GRAPH: Test Grade vs. Study Time

Line of Best Fit: A line placed on a graph that goes through the middle of the data points as evenly as possible. There is a formula for doing this and a line of best fit can be found using a graphing calculator. For our purposes, a sketch is adequate and it is sketched on the graph above. It doesn't matter if any data points are on the line or not. To write the equation of this line, use the point/slope formula on page 65.

Causation: A reason for why certain data relates or doesn't relate to the study. In the example above, the cause of higher grades does appear to relate to the hours of study time.

Correlation: Implies a relationship between the data and the study. It does not imply the reason for the relationship. There are three types of correlation that are easily determined by examining a scatter plot and its line of best fit.

1) Positive Trend - the line of best fit has a positive slope. The data points are plotted showing the general trend moving up as the graph is read from left to right.
2) Negative Trend - The line of best fit has a negative slope. The data points show a trend of going down as the graph is read from left to right.
3) No trend - no correlation. The data points are "all over the place."

Conclusion: This graph shows a positive correlation between test grade and time spent studying. The more time spent studying, the better the grade!

28 – ERROR IN MEASUREMENT

Actual (or True) Measure: The measure of a quantity that is accepted as being accurate and correct. This might be obtained from a chart, reference table, or other acceptable resource.

Experimental Measure: The measure of a quantity obtained by the person performing the measurement.

Example A student measures the mass of an object to be 156 grams. The actual mass is 173 grams.

– Actual Measure is 173 grams

– Experimental Measure is 156 grams

Absolute Error: The actual physical difference between the actual measure and the experimental measure of the item. The absolute error is the absolute value of the difference between the actual measure and the experimental measurement. It is a positive number and is labeled with whatever units are used in the measure.

– Actual Measure – Experimental Measure = $E_{absolute}$

– Absolute Error: $173\,g - 156\,g = 17g$

Relative Error: The relative error shows the importance of the error in relation to the size of the object measured. When we have an accepted value given in the problem, use this formula to find the relative error:

$$E_{relative} = \frac{E_{absolute}}{M_{actual}} \qquad\qquad E_{relative} = \frac{17\,g}{173\,g} = 0.098265896$$

We rounded this number to the nearest 1000th for convenience. $E_{relative} = 0.098$

Percent of Error: Often the relative error is expressed in % form.

$$0.098 = 9.8\%$$

Ratio of Absolute Error to Actual Measure in different circumstances:
The quantity of the absolute error has different meaning in problems with various actual measurements.

In another experiment, the absolute error might again be 17 grams just as it was in our primary example. However, if the actual measure is smaller than the example above, the relative error and percent of error would be considerably higher.

Example If the actual measure is 75 g, the relative error is $\dfrac{17g}{75g} \approx 0.227$ and the percent of error would be 22.7%.

On the other hand, if the actual measure is larger than the 173 g in our original experiment, the relative error is smaller and the percent of error is less.

Example If the actual measure is 300 g, the relative error is $\dfrac{17g}{300g} \approx 0.057$ which is only 5.7% of error.

Examining error in problem solving.
When quantities that involve a measurement error are used to calculate another answer, an even larger error can result. So when a measurement with a small relative error is used in a subsequent calculation, the final answer often involves a much larger relative error.

Example Using a ruler, Graeme measured the sides of a rectangular prism. His measurements were L = 5.2 cm, W = 8.3 cm, and the height was 4.2 cm. The actual correct measurements of the rectangular prism were L = 5.0cm, W = 8.0cm, and H = 4.0.

Linear

Percent of error in Length = $\dfrac{5.2\,cm - 5.0\,cm}{5.0\,cm} = 0.04 = 4\%$

Percent error in Width = $\dfrac{8.3\,cm - 8.0\,cm}{8.0\,cm} = 0.0375 = 3.75\%$

Percent error in Height = $\dfrac{4.2\,cm - 4.0\,cm}{4.0\,cm} = 0.05 = 5\%$

> The average percent of error for the 3 edges is 4.25%.

Area

Area of one side using Graeme's measurement:
$$A = LW$$
$$A = (5.2)(8.3) = 43.16 \text{ cm}^2$$

Area using correct measurements:
$$A = LW$$
$$A = (5)(8) = 40 \text{ cm}^2$$

Absolute error: $43.16 - 40 \text{ cm} = 3.16 \text{ cm}$

Relative Errorr: $\dfrac{3.16 \text{ cm}}{40 \text{ cm}} = .079$

Percent of error: 7.9%

Notice that percent of error for the area is larger than the percent of error in the linear measurement.

Volume

Volume of the rectangular solid using Graeme's measurements:
$$V = LWH$$
$$V = (5.2)(8.3)(4.2) = 181.272 \text{ cm}^3$$

Volume using correct measurements
$$V = LWH$$
$$V = (5.0)(8.0)(4.0) = 160.0 \text{ cm}^3$$

Percent of Error in Volume $\dfrac{181.272 - 160}{180} = 0.13295 = 13.295\%$

Conclusion: Although the original percent of error in measurement may not seem high, when the measurements are used in the area formulas, the percent of error increases. When they are used in the volume, the percent of error is significantly higher, more than tripled in this particular example!

PROBLEM SOLVING STRATEGIES

There are many different ways to solve problems. Although algebra is often used to solve math problems, nearly every single problem can be solved in several ways. It is important for students to describe the strategy they are using when they are working. Some problems may require more than one strategy. An organized method of showing your work makes problem solutions understandable to others. Some ideas to use in problem solving are:

- Work backwards.
- Make an equation.
- Guess & Check intelligently.
- Organize the data.
- Use logical reasoning.

- Find a pattern.
- Adopt a different point of view
- Consider all the possibilities.
- Consider extreme cases.
- Solve a simpler problem that is similar.

- Make a diagram, chart, map, or other visual representation.

Examples

❶ Express the sum of these fractions as a single fraction in simplest form.

$$\frac{1}{2x} + \frac{1}{x+3}$$ If you are unsure how to solve this, try solving an easier problem.

<u>Similar but easier problem</u>:

If $x = 2$, then $$\frac{1}{2(2)} + \frac{1}{(2)+3} = \frac{1}{4} + \frac{1}{5} = \frac{9}{20}$$

If $x = 3$, then $$\frac{1}{2(3)} + \frac{1}{(3)+3} = \frac{1}{6} + \frac{1}{6} = \frac{2}{6} = \frac{1}{3}$$

If $x = 4$, then $$\frac{1}{2(4)} + \frac{1}{(4)+3} = \frac{1}{8} + \frac{1}{7} = \frac{15}{56}$$

As you do the substitution, a **pattern develops**. The numerator of the sum is the sum of the denominators, and the denominator of the sum is the product of the denominators of the original fractions. Now apply the pattern to the algebraic fraction:

$$\frac{1}{2x} + \frac{1}{x+3} = \frac{2x+x+3}{2x(x+3)} = \frac{3x+3}{2x^2+6x}$$

Integrated Algebra Made Easy

❷ Oksana spent 1/2 of her money on a sweater. Then she spent 1/3 of what was left on makeup. She has $12.00 left. How much money did she have to start with? (2 strategies are shown)

Steps:

1) **Work backwards:** The $12.00 she has left over is 2/3 of the half she had after she bought the sweater. That means she had $18.00 after she bought the sweater. If $18.00 is the half that she had left, then she started with twice that amount, or $36.00. (The sweater was $18.00, the makeup was $6.00, and she still has $12.00.)

2) **Make a diagram:** Draw a circle to represent the total money she started with. Cut it in half and label one half "sweater". Draw a line to show 1/3 of the other half. This represents what she spent on makeup. The other 2/3 of that half is the money left which is $12.00. Working backwards, she spent $6 on makeup, and $18.00 on the sweater. Add up all the sections to get the total amount of $36.00.

❸ In a random drawing, find the probability of drawing a red marble or a green marble from a bag containing 3 red, 2 blue, and 5 green marbles.
Usual approach: P(R) + P(G) = 3/10 + 5/10 = 8/10

Different point of view: If we want red or green, then we don't want blue! Find the probability of drawing a blue marble and then subtract that from 1 to find the probability of "not blue": P(B) = 2/10. P(~B)=1–2/10 = 8/10

Colton has half as many dimes as quarters. How many of each does he have if he has $7.20 altogether? (2 strategies are shown.)

1st Method

Make a Chart and Guess Intelligently

# of Dimes	$ Value	# of Qtrs	$ Value	Total Value in $
8	$0.80	16	$4.00	$4.80 Not enough
10	$1.00	20	$5.00	$6.00 Better, but not yet
12	$1.20	24	$6.00	$7.20 Perfect!
14	$1.40	28	$7.50	$8.90 Too much!

Note: It is necessary to show a minimum of 3 "guess and checks" even if the first trial or guess works and checks. The correct guess must have a guess that is too small and another that is too large also shown. Make sure the right answer is clearly indicated as well!

Conclusion: Colton has 12 dimes and 24 quarters.

2nd Method

Make an Equation: Let x = # of dimes and let $2x$ = # of quarters.
$0.10x + 0.25(2x) = 7.20$
Solve for x and then find $2x$.
The conclusion is the same as shown above.

Guess & Check

Since calculators are so comfortably in use, students can often find the answers to problems using "trial and error" or "guess and check." In the Instructional Recommendations for High School Mathematics Instruction, published by The University of the State of New York, information about trial and error or "guess and check" is given as follows.

"Students who use trial and error to solve a problem must show their method. Merely showing that the answer checks or is correct is not considered a complete response for full credit. In order to receive full credit the student must show more than two trials with appropriate checks. The student will receive full credit if one of the following trial-and-error methods is used.

1. The student shows a developing pattern that progresses towards the solution, correctly identifies the solution, and provides appropriate checks.

2. The student "guesses" the solution with the first guess, provides an incorrect answer less than the solution and an incorrect answer greater than the solution with appropriate checks."

Integrated Algebra Made Easy

❺ In Claire's school there are 200 total students. Out of all the students, 70 play soccer, 55 play baseball, and 30 play tennis. There are some students who play two sports. They are as follows: 30 students play only soccer and baseball, 15 play only baseball and tennis, and 10 play only soccer and tennis. There are only 5 students who play all three sports. How many students do not play any of these sports?

Make a visual representation: Use a Venn Diagram. The overlapping parts of the appropriate circles contain the number of students who play two or three sports. Here is the breakdown.

• Soccer: Total of 70 students

Out of the 70 students, 30 play soccer and baseball, 10 play soccer and tennis, and 5 play all three sports. That is a total of 45 children, which leaves 25 that play only soccer.

• Baseball: Total of 55 students

Out of 55 students, 15 play baseball and tennis, 30 play soccer and baseball, and 5 students play all three sports. That leaves 5 students who play baseball only.

• Tennis: Total of 30 students

Out of 30 students, 15 play tennis and baseball, 10 play tennis and soccer, and 5 play all three sports. That is a total of 30 students. Therefore, there are 0 students who play tennis only.

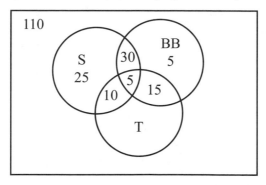

Conclusion: Add all the sections together to equal 90 students. that leaves 110 students who do not play any of these sports.

New York State Mathematics Glossary

This Glossary provides an understanding of the mathematical terms used in the Regents-approved course entitled Integrated Algebra as reflected in the New York State Mathematics Core Curriculum (Revised 2005). We encourage all students to become familiar with these terms .

A

absolute value: The distance from 0 to a number n on a number line. The absolute value of a number n is indicated by $|n|$.

absolute value function: A function containing the absolute function of a variable.

accuracy: How close a numerical measure is to its actual value.

acute angle: An angle whose measure is greater than $0°$ and less than $90°$.

adjacent angles: Two coplanar angles that share a common vertex and a common side but have no common interior points.

adjacent sides: Two sides of any polygon that share a common vertex.

adjacent side of an acute angle in a right triangle: The leg of the right triangle that is a side of the acute angle.

algebra: The branch of mathematics that uses letters and/or symbols, to represent numbers and express mathematical relationships.

algebraic equation: A mathematical statement that is written using one or more variables and constants which contains an equal sign.

algebraic expression: A mathematical phrase that is written using one or more variables and constants, but which does not contain a relation symbol $(<, >, \leq, \geq, =, \neq)$

algebraic fraction: A fraction that contains an algebraic expression in its numerator and/or denominator.

algebraic representation: The use of an equation or algebraic expression to model

algorithm: a defined series of steps for carrying out a computation or process.

angle of depression: The angle formed by the horizontal and the line of sight when looking downward.

angle of elevation: The angle formed by the horizontal and the line of sight when looking upward.

appropriateness: Reasonableness of an answer or method.

approximate value: A value for some quantity, accurate to a specified degree.

argument: The communication, in verbal or written form, of the reasoning process that leads to a valid conclusion.

array: A set of objects or numbers arranged in an order, usually in rows and columns.

associative property: A property of real numbers that states that the sum or product of a set of numbers or variables has the same value, regardless of how the numbers or variables are grouped.

axis: A horizontal or vertical line used in the Cartesian coordinate system used to locate a point.

B

base: A number or an expression that is raised to a power.

binomial: An algebraic expression consisting of two terms.

bivariate data: Data involving two variables.

box-and-whisker plot: A visual display of a set of data showing the five number summary: minimum, first quartile, median, third quartile, and maximum. This plot shows the range of scores within each quarter of the data. It is very useful for examining the variation in a set of data and comparing the variation of more than one set of data.

C

circle: The set of all points (or locus of points) in a plane that are a fixed distance, (called the radius) from a fixed point, (called the center).

closure: A set "S" and a binary operation "*" are said to exhibit closure if applying the binary operation to any two elements in "S" produces a value that is a member of "S".

coefficient: The numerical factor of a term in a polynomial.

common base(s): Exponential expressions or equations that have the same or equivalent bases.

---

common factor: A number, polynomial, or quantity that divides two or more numbers or algebraic expressions evenly.

commutative property: A property of real numbers that states that the sum or product of two terms is unaffected by the order in which the terms are added or multiplied; i.e., the sum or product remains the same.

compare: To state the similarities or differences between two or more numbers, objects, or figures by considering attributes such as size, shape, odd, even.

complement of a set: The elements of a universe not contained in a given set; the subset that must be added to any given subset to yield the original set. The complement of set A is indicated by A' or A^C.

Universe - interior of the square

Set A = the circular region

Complement of A is A' (or A^c)

conditional probability: A probability that is computed based on the assumption that some event has already occurred. The probability of event B given that event A has occurred is written P (B|A).

conjecture: An educated guess; an unproven hypothesis based on observation, experimentation, data collection, etc.

coordinate: An ordered pair of numbers that identifies a point on a coordinate plane, written as (x, y). The number represented by "x" is called the x-coordinate (abscissa). The number represented by "y" is called the y-coordinate (ordinate).

correlation: A statistical measure that quantifies how pairs of variables are related; a linear relationship between two variables.

cosine: For a given acute angle θ in a right triangle, the ratio of the length of the side adjacent to an acute angle to the length of the hypotenuse. The cosine of an angle is written as COS. See also circular function.

cubic unit: A unit for measuring volume.

cumulative frequency table: A table that shows how often each item, number, or range of numbers occurs in a set of data. This table displays the total number of scores that fall into each of several cumulative intervals. The cumulative intervals are created by adding the preceding tallies (of lower scores) to the new tallies for each interval.

---

cylinder: A solid geometric figure bounded by two parallel bases which are congruent circles and a lateral surface which consists of the union of all line segments joining points on each of those circles.

D

decagon: A polygon with ten sides.

degree of a monomial: The sum of the exponents of the variables in the monomial.

degree of a polynomial: The highest degree of any monomial term in the polynomial.

dependent events: Two events in which the outcome of the first event affects the outcome of the second event

dependent variable: A variable whose value is determined by a second variable.

difference of two perfect squares: A binomial of the form $a^2 - b^2$ *which can be factored into* $(a - b)(a + b)$.

distributive property: A property of real numbers that states that the product of a number and the sum or difference of two numbers is the same as the sum or difference of their products.

E

element: An object contained in a set.

empirical probability: An estimate of the probability of an event based on the results of repeated trials of the event.

equation: A mathematical sentence stating that two expressions are equal.

exponent: A number that tells how many times the base is used as a factor of a term; in an expression of the form b^n, n is called the exponent, b is the base, and b^n is a power of b.

exponential decay: The decreasing exponentially of a quantity over time represented by $y = a \cdot b^x$ where $a > 0$ and $0 < b < 1$.

exponential form: An expression or equation containing exponents.

exponential function: A function with a variable in the exponent; an equation in the form $y = ab^x$, where $a \neq 0$ and $b > 0$, $b \neq 1$.

---

exponential growth: The increasing exponentially of a quantity over time represented by $y = a \cdot b^x$ where $a > 0$ and $b > 1$.

expression: A mathematical representation containing numbers, variables, and operation symbols; an expression does not include an equality or inequality symbol.

extrapolate: The process of using a given data set to estimate the value of a function or measurement beyond the values already known.

F

factor: *(noun)* A whole number that is a divisor of another number; an algebraic expression that is a divisor of another algebraic expression.

factor: *(verb)* Find the number of algebraic expressions that give an indicated product.

five number summary: For a data set, these include the minimum, the first quartile, the median, the third quartile, and the maximum.

frequency table: A table that shows how often each item, number, or range of numbers occurs in a set of data.

function: A rule that assigns to each number x in the function's domain a unique number $f(x)$.

G

geometry: Branch of mathematics that deals with the properties, measurement, and relationships of points, lines, angles, surfaces, and solids.

graphical representation: A graph or graphs used to model a mathematical relationship.

graphical solution of a system of equations: The set of points in the plane whose coordinates are solutions to a system of equations.

greatest common factor (GCF): The greatest number or expression that is a factor of two or more numbers or expressions.

H

hexagon: A polygon with six sides.

histogram: A frequency distribution for continuous quantitative data. The horizontal axis is a number line that displays the data in equal intervals. The frequency of each bar is shown on the vertical axis.

---

hypotenuse: The side of a right triangle opposite the right angle; the longest side of a right triangle.

I

identities: Equations that are true for all values of the variables they contain.

identity elements: For a binary operation * and a set S, I is the identity element if $a * I = a$ and $I * a = a$ for every element a that is in S.

image: The resulting point or set of points under a given transformation; in any function f, the image of x is the functional value $f(x)$ corresponding to x.

impossible event/outcome: An event that cannot occur. The probability of an impossible event equals zero.

independent events: Two or more events in which the outcome of one event has no effect on the outcome of any other event.

independent variable: An element in the domain of a function; the input value of a function.

inductive reasoning: The process of observing data, recognizing patterns and making generalizations about those patterns.

inequality: A mathematical statement containing one of the symbols $<, >, \geq, \leq, \neq$ to indicate the relationship between two quantities.

integers: The set of numbers that is the union of the counting numbers, their opposites, and zero (i.e.,$\{...-4, -3, -2, -1, 0, 1, 2, 3, 4, ... \}$).

interpolate: The process of using a given data set to estimate the value of a function or measurement between the values already known.

interquartile range: The difference between the first and third quartiles; a measure of variability resistant to outliers.

intersection of sets: The intersection of two or more sets is the set of all elements that are common to all of the given sets.

inverse operation: An operation that undoes another operation; addition and subtraction are inverse operations; multiplication and division are inverse operations; raising to a power and taking a root are inverse operations.

J

There are no J terms in the commencement-level sections of the NYS Mathematics Core Curriculum (Revised March 2005).

K

There are no K terms in the commencement-level sections of the NYS Mathematics Core Curriculum (Revised March 2005).

L

leading coefficient: The coefficient of the first term of a polynomial when the polynomial is in standard form.

legs of a right triangle: The two sides of a right triangle that form the right angle.

like radical terms: Terms that have the same index and the same radicand.

line of best fit: A line used to approximate and generalize the linear relationship between the independent and dependent variables for a set of data. It may not be equivalent to a least squares regression model.

linear equation: A first degree equation.

linear inequality: An inequality of the first degree.

linear transformation: A transformation of data set X is of the form $X' = a + bX$, where a is the additive component and b is the multiplicative component.

literal equation: An equation that contains more than one variable.

logical argument: A reasoning process based on logic that uses a series of statements leading to a conclusion.

M

mean: A measure of central tendency denoted by x, read "x bar", that is calculated by adding the data values and then dividing the sum by the number of values. Also known as the arithmetic mean or arithmetic average.

measure of central tendency: A *summary statistic* that indicates the typical value or center of an organized data set. The three most common measures of central tendency are the mean, median, and mode.

median: A measure of central tendency that is, or indicates, the middle of a data set when the data values are arranged in ascending or descending order. If there is no middle number, the median is the average of the two middle numbers.

---

mode: A measure of central tendency that is given by the data value(s) that occur(s) most frequently in the data set.

monomial: A polynomial with one term; it is a number, a variable, or the product of a number (the coefficient) and one or more variables.

multiple representations: Various ways, i.e., graphically, numerically, algebraically, geometrically, and verbally, to present, interpret, communicate, and connect mathematical information and relationships.

multiplication property of zero: For every number a, $0 \cdot a = 0$ and $a \cdot 0 = 0$.

mutually exclusive events: Two events that cannot occur at the same time.

N

nonagon: A polygon with nine sides.

null set: The set with no elements. The empty set can be written $\varnothing$ or $\{\ \}$

O

octagon: A polygon with 8 sides.

opposite side in a right triangle: The side across from an angle. In a right triangle the hypotenuse is opposite the right angle and each leg is opposite one of the acute angles.

ordered pair: Two numbers that are used to identify the position of a point in a plane. The two numbers are called coordinates and are represented by (x, y).

ordinate: The vertical coordinate of a two-dimensional rectangular coordinate system; usually denoted by y.

P

parabola: The locus of points equidistant from a given point (called the focus) and a given line (called the directrix). A common form of an equation of a parabola with vertical line symmetry is where a, b, and c are real numbers and $a \neq 0$.

parallel lines: Two or more coplanar lines that do not intersect. Parallel line segments or rays are line segments or rays that are subsets of parallel lines.

parallelogram: A quadrilateral in which both pairs of opposite sides are parallel.

parameter: A quantity or constant whose value varies with the circumstances of its application.

pentagon: A polygon with 5 sides.

percent of increase/decrease: The magnitude of increase/decrease expressed as a percent of the original quantity.

perimeter: The sum of the lengths of all the sides of any polygon.

polygon: A closed plane figure formed by three or more line segments that meet only at their endpoints.

polynomial: A monomial or sum of monomials.

premise: A proposition upon which an argument is based or from which a conclusion is drawn.

prime factorization: Writing an integer as a product of powers of prime numbers.

probability: The likelihood of an event occurring. The probability of an event must be greater than or equal to 0 and less than or equal to 1.

product property of proportions: In a proportion $\frac{a}{b} = \frac{c}{d}$, the product of the means (b and c) equals the product of the extremes (a and d), or in other words: $b \cdot c = a \cdot d$.

proof: A logical argument that establishes the truth of a statement; a valid argument, expressed in written form, justified by axioms, definitions, and theorems.

properties of the real numbers: Rules that apply to the operations with real numbers.

proportional: Two variables are proportional if they maintain a constant ratio. See also direct variation.

Pythagorean theorem: The mathematical relationship stating that in any right triangle the sum of the squares of the lengths of the two legs is equal to the square of the length of the hypotenuse; if a and b are the lengths of the legs and c is the length of the hypotenuse, then $a^2 + b^2 = c^2$.

Q

quadratic equation: An equation that can be written in the form $ax^2 + bx + c = 0$, where a, b, and c are real constants and $a \neq 0$.

quadrilateral: A polygon with 4 sides.

quantitative: Descriptions using numerical measures such as quantity, height, or age.

R

radical: The root of a quantity as indicated by the radical sign.

radicand: The quantity under a radical sign; a *number or expression from which a root is extracted.*

range (of a data set): The difference between the maximum and minimum data values in a data set

rates: A ratio that compares quantities of different units (e.g., miles per hour, price per pound, students per class, heartbeats per minute).

ratio: A comparison of two quantities having same units (e.g., 2 to 3, 2:3, $\frac{2}{3}$).

rational coefficient: A coefficient that is a rational number.

rational expression: The quotient of two polynomials in the form $\frac{A}{B}$, where are A and B polynomials and $B \neq 0$.

rational number: Any number that can be expressed as a ratio in the form $\frac{a}{b}$ where a and b are integers and b. A rational number is either a terminating or repeating decimal.

real numbers: The set of numbers that includes all rational and irrational numbers.

rectangle: A parallelogram containing one right angle; a quadrilateral with four right angles.

rectangular coordinates: An ordered pair of real numbers that establishes the location of a point in a coordinate plane using the distances from two perpendicular intersecting lines called the coordinate axes. (See also Cartesian coordinates.)

rectangular solid: A prism whose six faces are rectangles.

regular polygon: A polygon which is both equilateral and equiangular.

relation: A correspondence between two sets; a set of ordered pairs

relative error: The ratio of the absolute error in a measurement to the size of the measurement; often written as a percent and called the percent of error; the absolute error is the difference between an approximation and the exact value.

representations: Models, (e.g., symbolic, verbal, graphical, numerical, physical, pictorial) used to represent and interpret mathematical problems.

rhombus: A parallelogram with two adjacent congruent sides; a quadrilateral with four congruent sides.

right angle: An angle formed by two perpendicular lines, the measure of which is 90°.

right triangle: A triangle with one right angle.

root of an equation: A solution to an equation of the form $f(x) = 0$.

roster form: A notation for listing all the elements in a set using set brackets and a comma between each element.

S

sample space: The set of all possible outcomes for a given event.

scatter plot: A graphical display of statistical data plotted as points on a coordinate plane to show the correlation between two quantities.

scientific notation: A convenient way to write very <u>small</u> or <u>large numbers</u>. In scientific notation, numbers are separated into two parts, a real number with an <u>absolute value</u> equal to or greater than 1 and less than 10 and an <u>order of magnitude</u> value written as a <u>power</u> of 10.

semi-circle: Either of the arcs of a circle determined by the endpoints of a diameter.

set: A well-defined collection of items.

set-builder notation: A notation used to describe the elements of a set.

simplest form: An expression that has been rewritten as simply as possible using the rules of arithmetic and algebra.

sine: For a given acute angle θ in a right triangle, sin θ, is the ratio of the length of the side opposite the acute angle θ to the length of the hypotenuse. See also circular function.

slope: The measure of the steepness of a line; the ratio of vertical change to horizontal change; if point P is (x_1, y_1) and point Q is (x_2, y_2) the slope of $\overline{PQ}$ is $\dfrac{\Delta y}{\Delta x} = \dfrac{y_2 - y_1}{x_2 - x_1}$.

solution set: Any and all value(s) of the variable(s) that satisfy an equation, inequality, system of equations, or system of inequalities.

square: A rectangle with two congruent adjacent sides.

square units: The basic unit of area.

---

subset: A set consisting of elements from a given set; it may be the empty set.

substitution property: Any quantity can be replaced by an equal quantity.

subtraction property of equality: If the same or equal quantities are subtracted from same or equal quantities, then the results are equal.

surface area: The sum of the areas of all the faces or curved surfaces of a solid figure.

system of equations/inequalities: A set of two or more equations/inequalities. The solution set contains those values that satisfy all of the equations/inequalities in the system.

T

tangent (of an angle): For a given acute angle θ in a right triangle, $\tan \theta$ is the ratio of the length of the side opposite the acute angle θ to the length of the side adjacent to the angle. See also circular function.

trapezoid: A quadrilateral with exactly one pair of parallel sides.

triangle: A polygon with three sides.

trigonometry: The branch of mathematics that deals with trigonometric functions.

trinomial: A polynomial with exactly three terms.

U

undefined: An expression in mathematics which does not have meaning and therefore is not assigned a value.

union of sets: The union of two or more sets is the set of all elements contained in at least one of the sets.

univariate: A set of data involving one variable.

universe: The set of all possible specified elements from which subsets are formed. Also know as the universal set.

V

valid argument: A logical argument supported by known facts or assumed axioms; an argument in which the premise leads to a conclusion.

variable: A quantity whose value can change or vary; in algebra, letters often represent variables.

Integrated Algebra Made Easy

Venn diagram: A drawing showing relationships among sets.

vertex of an angle: The point of intersection of the two rays that form the sides of the angle.

vertex of a polygon: A point where the edges of a polygon intersect.

volume: A measure of the number of cubic units needed to fill the space inside a solid figure.

visualization: A mental image based on a given description.

X

x-axis: One of the two intersecting lines used to establish the coordinates of points in the Cartesian plane; in that plane, the line whose equation is $y = 0$; in space the axis perpendicular to the yz-plane.

x-coordinate: The first coordinate in any (x, y) ordered pair; the number represents how many units the point is located to the left or right of the y-axis; also called abscissa.

x-intercept: The point at which the graph of a relation intercepts the x-axis. The ordered pair for this point has a value of $y = 0$.

Y

y-axis: One of the two intersecting lines used to establish the coordinates of points in the Cartesian plane; in that plane, the line whose equation is $x = 0$; in space the axis perpendicular to the xz-plane.

y-coordinate: The second coordinate in any (x, y) ordered pair; the number represents how many units the point is located above or below of the x-axis; also called ordinate.

y-intercept: The point at which a graph of a relation intercepts the y-axis. The ordered pair for this point has a value of $x = 0$.

Z

z-coordinate: The third coordinate in any (x, y, z) ordered triple; the number represents how many units the point is located above or below the xy-plane.

zero product property: If a and b are real numbers, then $ab = 0$ if and only if $a = 0$ or $b = 0$, or a and $b = 0$.

INDEX

A

Absolute Value 13, 27, 81
Addition 18, 59
  Fractions 36
Age Problems 57
Algebraic Expression 26
Algebraic Fraction 21, 34-40
  Simplify or Reducing 34
Algebraic Solution 58, 90
  Substitution 58
  Addition 59
Area 99, 104
Associative Property 7, 12
Axis of Symmetry 87

B

Biased 120
Binary Operation 10
Binomial 17
Bivariate Data 120
Box and Whisker Plot 125

C

Causation 126
Certainty 107
Circle 99, 104
Circumference 104
Closure 11
Coefficient 42
Coin Problems 56
Collinear Points 64, 68
Combinations 118
Commutative Property 7, 12
Complement 1, 107
Completing The Square 86
Consecutive Integer 55

Constant 17
Conversion 51
Conversion Factors 50

Coordinate Graph 63
Correlation 126
Cosine 93
Counter-Example 12
Counting Principle 114
Cumulative Frequency 121, 114
Cylinders 100, 103

D

Data 120
Decimal Equations 54
Degree of Polynomials 23
Dependent Variable 64
Direct Variation 49
Discount 45
Distributive Property 8, 12
Division 15, 19, 20
  by Zero 15
  Fractions 15, 38
  Long 21, 121

E

Equal Chance 109, 111
Equation 17
  Graphing 63, 73, 88
  Line 70
  Quadratic 84, 88, 89
  Simple 53
  Solving 53
  Decimal 54
  Fractional 54
  Literal 54
Equivalent Fractions 36
Evaluate 26
Exponent 22
  Negative 23
  Zero 23
Expression(s) 17, 26-27
  Algebraic 26
  Exponential 27

EXACT Answers 42
Events
   Independent 106, 109, 110
   Dependent 106, 111, 112

F

F.O.I.L. 19
Factorials 27, 114, 115
Factoring 28-33
   Binomials 29-33
   Trinomials 29-33
Fractions
   Algebraic 21, 34
   w/ Signs 36
   Equivalent 36
   Addition 36
   Subtraction 36
   Multiplication 38
   Division 15, 20, 38
   Actions Undefined 21
Fractional Equation 54
Frequency Table 121, 123
Function 78
   Linear 79
   Quadratic 80
   Absolute Value 81
   Exponential 82

G

Geometry Problems 46
Geometry Word Problems 57
Graphing 58, 121
   Coordinates 63
   Inequalities 62
   Functions and Relations 78
   Line 72
   Quadratic Equations 88-89
   Parabola 89
   Systems of Inequalities 75-77
   Systems of Linear Equations 58, 75

Greatest Common Factor 20, 29

H

Histogram 122, 124
Horizontal Axis (x-axis) 63

I

Identity 9
   Additive 9
   Multiplicative 9
Increase 46
Independent Variable 64
Index 41
Inequality 49
   Solving 61-62
   Graphing 62
Integer 5, 55
   Positive or Negative 55
   Consecutive 55
   Odd or Even 55
Intervals 121
Inverse 9
   Additive 9
   Multiplicative 9
Irrational Numbers 6

L

Least Common Denominator 36
Like Terms 17
Line
   Best Fit 126
Linear Equation 58
Linear Function 79
Literal Equation 54
Long Division 21

M

Mean 121-125
Measurement Error 127
Measurement System
  English 51
  Metric 51
Median 121-125
Metric 51
Mode 121-125
Monomials 17-21
Multiplication 18
  Fractions 38

N

Natural Numbers 5
Number Line 3

O

Operation 10
OR 113
Order of Operations 15
Ordered Pair 63
Origin 63
Outcomes 106, 107, 13
Outlier 122

P

Parabola 88
Parallel Lines 69
Parallelogram 98
Percentiles 122-125
Perimeter 99
Permutations 115-119
Perpendicular
  Lines 69
Pi 6, 105
Polyhedron 100
Polynomials 17-21
Powers 22
  Of Ten 24, 25

Prime Factors 28
Prism 100
Probability 106-115
Proportion 40, 44, 47, 49
Pythagorean Theorem 92
Pythagorean Triples 93

Q

Quadrant 63
Quadratic 84
Quadratic - Linear Pair 90-91
Quadratic Equations Graphically 87-89
Quadratic Formula 86
Quadratic Functions 80
Quadratic Word Problems 86
Quadrilaterals 97, 98
Qualitative Variable 120
Quantitative Variable 120
Quartiles 122-125

R

Radicals 41-43
  Adding or Subtracting 43
  Multiplying or Dividing 43
  Square Root 41
  Simplifying 42
Radicand 41
Radius 104
Range 121
Ratio 44
  Word Problems 47, 56
Ratio Problems 56
Rational Numbers 5
Rates 51-53
Real Numbers 4, 9
Reciprocal 4
Regular Polygon 100
Relation 78
Right Triangle 92-96
Root 41, 84, 87

S

Sample Space 107, 13
Scale Drawing 48
Scatter Plot 126
Scientific Notation 24
Simplifying Radicals 42
Similarity 46, 47
Sine 93
Slope 64-67
  Equals Zero 67
  Formula 65
  Intercept 68, 73
  Intercept Form 64
  Undefined 67
SOHCAHTOA 93
Special Right Triangles 96
Square Root 41, 86
Standard Form 23
Statistics 120
Subtraction 8, 14, 18, 36
Simplifying 34

T

Table of Values 74, 88
Tally Sheet 121
Tangent 93
Term 17
Tree Diagram 107-112
Triangle
  Special Right 96
Trigonometry, Right Triangle 93
Trinomial 17
Turning Point 87

U

Unbiased 120

Unequal Chance 110, 112
Univariate Data 120
Unlike Terms 17

V

Variable
  2 Variable Word Problems 60
  Independent 64
  Dependent 64
  Qualitative 120
  Quantitative 120
Venn Diagram 131
Vertex 87
Vertical Axis (y-axis) 63
Volume of Solids 100

W

Whole Numbers 5
Word Problem 55
Word Problems with Ratios 47
Word Problems with 2 Variables 60

Y

Y-Intercept 68, 70

Z

Also Check Out

Other books in this series

Geometry Made Easy Handbook
ISBN #978-1-929099-37-5

Algebra 2/Trigonometry Made Easy Handbook
ISBN #978-1-929099-91-7

Written by: MaryAnn Casey

AVAILABLE ONLY AT THE

TOPICAL REVIEW BOOK COMPANY

PHONE: 1-800-847-0854 • FAX: 1-800-847-0851

E-MAIL: topiclarbc@aol.com • WEBSITE: www.topicalrbc.com